Primate People

primate
PEOPLE

SAVING NONHUMAN PRIMATES
THROUGH EDUCATION, ADVOCACY, & SANCTUARY

EDITED BY Lisa Kemmerer

FOREWORD BY Marc Bekoff

THE UNIVERSITY OF UTAH PRESS
Salt Lake City

The Defiance House Man colophon is a registered trademark
of the University of Utah Press. It is based upon a four-foot-tall,
Ancient Puebloan pictograph (late PIII) near Glen Canyon, Utah.

16 15 14 13 12   1 2 3 4 5

LIBRARY OF CONGRESS CATALOGING-IN-PUBLICATION DATA
Primate people : saving nonhuman primates through
education advocacy and sanctuary /
   edited by Lisa Kemmerer.
      p. cm.
Includes bibliographical references and index.

  ISBN 978-1-60781-178-7 (cloth : alk. paper)
  ISBN 978-1-60781-153-4 (pbk. : alk. paper)
  ISBN 978-1-60781-215-9 (ebook)

    1. Primates—Conservation—Anecdotes.
    2. Primates—Research—Anecdotes.
    3. Primatologists—Anecdotes.
    4. Wildlife conservationists—Anecdotes. I. Kemmerer, Lisa.
QL737.P9P67283 2011
639.97'98—dc23

                              2011030174

Printed and bound by Sheridan Books, Inc., Ann Arbor, Michigan.
Frontispiece: Courtney, a gibbon. Photo courtesy of Keri Cairns.

This book is dedicated to people everywhere

who are doing something—anything—on behalf of nonhuman primates.
I send this book into the world with the hope that every single nonhuman primate—
every living being—may one day be as cherished and protected
as gibbons are in the care of Shirley McGreal.

Contents

Foreword

Nonhuman primates are in a great deal of danger, mainly because of the invasive ways of human primates. Numerous species are in peril, and time is not on their side—or ours—in this human-dominated world. It is thus essential that humans take serious and immediate action to assist these threatened species.

Molecular evidence indicates that, some eight million years ago, gorillas branched off the evolutionary line that eventually produced human beings. About a million years later, an unidentified primate branched off the gorilla line. Both chimpanzees and human beings descended from this vanished primate (Sagan and Druyan 1992, 343). This means that the closest human relative proves to be the chimp and the closest relative of the chimp is the human. Not orangutans but *Homo sapiens*. Us. Chimps and humans are nearer kin than are chimps and gorillas or any other kind of ape of another species. Gorillas are the next closest relatives, both to chimps and humans. By these standards, humans and chimps are about as closely related as horses and donkeys and are closer than mice and rats, or turkeys and chickens, or camels and llamas (Sagan and Druyan 1992, 277).

Consequently, chimpanzee DNA (including bonobos) is almost indistinguishable from human DNA. Human and chimpanzee DNA differs by 1.7 percent, and if we only consider the working genes, we differ by a mere 0.4 percent. Chimpanzees are genetically closer to us than are any other animals. Humans and gorillas differ by a mere 1.8 percent. Humans and orangutans vary by 3.3

percent, gibbons by 4.3 percent, humans and rhesus monkeys by 7 percent (Sagan and Druyan 1992, 276–77).

Our relationships with nonhumans are complicated, frustrating, ambiguous, and paradoxical. It should not matter how close our genetic code is to other creatures to treat them with respect—to protect and preserve their lives and well-being. It is this type of thinking—"if you are like me, then I will protect you"—that lies behind racism. Furthermore, caring about others—no matter how different from ourselves—does not allow unjust exploitation, no matter how great the gains. When people tell me that they love animals and then harm or kill them, I tell them I'm glad they don't love me. We observe animals, gawk at them in wonder, experiment on them, eat them, wear them, write about them, draw and paint them, and move them from here to there as we "redecorate nature." Respect and care for others is not consistent with taking advantage of others.

Primate People, a timely and most welcomed book, highlights the lives and misfortunes of our closest relatives as experienced by people working in sanctuaries and rehabilitation facilities, who describe these lively, but often damaged, individuals; their peculiar habits, silly antics, and affections; and the reasons they end up living in these places. This collection of essays offers a haunting vision of what happens to macaques and rhesus monkeys behind the closed doors of primate research facilities, written by people who have entered these places either before they knew what to expect or as undercover agents. *Primate People* also gives voice to people in Africa, Asia, South and North America, and Europe who work as veterinarians, sanctuary founders, grassroots activists, researchers, and lobbyists.

For all the misery that nonhuman primates have suffered and continue to suffer at the hands of humanity, *Primate People* is a forward-looking book of hope. Authors in this anthology show new understanding and a commitment to change; they display a compassion and dedication that is central to animal advocacy. Once readers understand the pressing problems that face nonhuman primates, perhaps they will avoid palm oil, volunteer at primate sanctuaries, and/or donate money to the International Primate Protection League because the most important truth about primates—about the suffering and need of all nonhumans—is that there is something each of us can do to help these relatives of ours and the countless other suffering individuals of innumerable other species. As I point out in my book, *The Animal Manifesto: Six Reasons for Expanding Our*

Compassion Footprint, it's easy to effect changes that will make the world a more peaceful place for nonhumans and us. Our own well-being depends on the fate of other species.

We are living in a time when the suffering of billions of pigs and monkeys seems of little concern to the masses of humanity. But a handful of people are working to pull back the curtain of sorrow, to remind the rest of us that we are the authors of our own destiny, the creators of our own civilizations; we must rethink our common assumptions and habitual practices, and, in the case of primates, we must do so quickly. Indeed, we need to move fast to stop our intrusions into the lives of billions of animals.

It's really pretty simple. This animal manifesto—"treat us better or leave us alone"—asks people to regard other animals as fellow sentient, emotional beings, to recognize the cruelty that too often defines our relationship with them, and to change our behavior so that we begin to act compassionately on their behalf.

Among my worst fears is that I'll wake up one day and wonder where all the animals have gone. Even if we do make immediate changes, numerous animals will perish. But that's where persistence and hard work come in. We don't have to produce a movie, write a book, or found a movement or organization to make a difference. Right now each of us can start making more humane and ethical choices in our daily lives—in the food we eat, the clothes we wear, the products we buy, and the cars we drive, to offer just a few examples. It doesn't take a great deal of effort to make a positive difference.

In contrast with our carbon footprint, our compassion footprint is something we need to enlarge. When we try to externalize our innate compassion through interacting with other living beings, we are making progress and expanding our compassion footprint. I'm an optimist and a dreamer, and I think that the future can be much better for animals—nonhuman and human—if those who have the ability are also willing to make changes to create a more compassionate and peaceful world.

Joel Cohen, head of the Laboratory of Populations at Rockefeller University and Columbia University, offers the following sobering fact: the difference in the population between less-developed areas of the world (the have-nots) and more developed regions (the haves) will increase from twofold in the 1950s to about sixfold by 2050. In the future, it is likely that fewer people will have the necessary resources to make a positive difference in our relationships with animals and ecosystems.

There's no room for hypocrisy or negativity. We need to put these aside and move ahead together as a tight-knit community, aware of how much work there is to do. We also need to step forward knowing and feeling that, with a global commitment to compassion and empathy, we can succeed. Never say never—ever. We need to rewild our hearts and build corridors of compassion and coexistence that include all beings. We're not the only show in town. In any event, we suffer the same indignities we impose on other beings.

Let's all work together to make this the Century of Compassion, the Era of Empathy. We need to treat animals better or make sure we stay out of their way. If we are unsure what this means, we can take our example from others. *Primate People* is a significant move in the right direction.

Marc Bekoff
BOULDER, COLORADO

Acknowledgments

Special thanks to Shirley McGreal for more than a quarter century of work on behalf of nonhuman primates, including her support for many other worthy organizations that try to save and protect them. Also many thanks to each author who submitted an essay and to Marc Bekoff for his willingness to write a foreword for this anthology.

Note to Readers

Please support the International Primate Protection League (IPPL)

Dr. Jane Goodall noted that many more people care about nonhuman primates "because of the awareness IPPL has raised in countries around the world" ("Awards" 2008, 3). IPPL's founder, Shirley McGreal, recently received the prestigious Order of the British Empire and was honored by the National Anti-Vivisection Society for her work on behalf of primates. You can read about IPPL and view its archived magazines (dating back to 1974) at http://www.ippl.org.

IPPL works across international boundaries to support worthy organizations that help nonhuman primates, including sanctuaries, rehabilitation and educational facilities, and grassroots organizations involved in political advocacy and habitat protection. IPPL staff members are up-to-date and informed on primate problems and solutions worldwide and can therefore disperse necessary funds to appropriate facilities wisely. IPPL donors can feel confident that their funds will assist worthwhile primate protection, rescue, and rehabilitation in facilities around the world.

Introduction

Lisa Kemmerer

A new species of Titi monkey was discovered and recorded in western Bolivia in 2004 (Madidi Titi), joining more than four hundred other recorded primate species. What do most of us know about these many other primates—about the proboscis monkey, Hainan gibbon, pygmy tarsier, or slow loris?

Earth was once rich with primates, but each species—except one—is now endangered because of just one primate: *Homo sapiens*. Meat industries threaten both South American and African primates. Roughly one hundred primate species live in Brazil, a nation where rain forests are leveled to pasture cattle in order to export meat to wealthy consumers. Africa's bushmeat trade (trade in non-human primate flesh) has been augmented by logging roads that wind deep into once isolated habitat. West Africa's bonobos, perhaps our closest relatives, have been devastated by the bushmeat trade: "In one human generation, 90 percent of the Bonobos have disappeared" (Brown 2008, 102). Miss Waldron's red colobus monkeys (also of West Africa) have been driven to probable extinction by the human appetite for their flesh.

Who were these primates, and what forces destroyed their existence? How many people outside of West Africa knew of Miss Waldron's red colobus monkeys? Would we care more about these individuals if we knew something of their lives and their suffering? Will we change what we purchase if we learn that our consumer choices harm and endanger other primates? Once informed, might we support those who work on behalf of endangered primates, or lobby for change ourselves? If we do, can we save Earth's many and wondrous nonhuman primates from ongoing, extreme suffering and the finality of extinction? My hope is that the answer is, "Yes."

Delacour's langurs cling to cliffs and sleep in caves on the borders of Vietnam, Laos, and China. Adults are largely black with a white band around their hips (the reason they are sometimes called "white shorts" monkeys) while their infants are distinctively orange. Delacour's langurs are small with bushy tails that reach nearly three feet—much longer than their bodies. These primates eat mostly leaves but may also dine on fruit, flowers, and seeds. They live in communities of eight or nine with only one adult male, serving as protector, in each group.

Unfortunately, agricultural development and quarrying have devastated Delacour's langur habitat; deforestation now separates diminished populations. Estimates indicate that there are only 317 of these monkeys left on the planet (Nguyen 2009, 4–5). Nonetheless, human predators continue to shoot these langurs from their sandstone-cliff homes to use their body parts for medicines.

Apes, in contrast to monkeys, have rotary shoulder blades, no tails, and proportionally larger brains. How much do most people know about gibbons, the smallest and most diverse species of ape? Gibbons are covered in thick, soft, woolly hair (Crair 2008, 6). Dozens of species of gibbons live high in the forest canopies of Southeast Asia ("Highland Farm" 2009, 6). They are the only apes who reside in the tops of trees. Gibbons swing on ball-and-socket wrists that rotate as much as 180 degrees. Swinging to and fro on their long, graceful arms, aided by unusually long fingers and nonopposable thumbs, they race across the forest canopy, reaching an astonishing thirty-five miles per hour.

When earthbound, gibbons walk on two legs and have extremely sophisticated vocalizations—they sing beautifully. Gibbons are territorial and live in monogamous couple relationships (Crair 2008, 7). Their haunting duets reinforce monogamous bonds and establish individual territories ("Wildlife Friends" 2009, 9).

Gibbons are at greater risk of extinction than any other ape (Crair 2008, 7). Although all gibbons are endangered, the Hainan gibbon of China is the most threatened: it is estimated that just twenty individuals remain ("Highland Farm" 2009, 6). The eastern black gibbon of Vietnam has been reduced to only a few dozen individuals. The Javan gibbon is also very rare, as is the western hoolock gibbon, whose population dropped from one hundred thousand to just five thousand in a handful of decades (Crair 2008, 6). Yet humans continue to shoot gibbons to sell their flesh and in the hope of capturing their babies, whom they sell into the pet and tourist industries ("Highland Farm" 2009, 6).

The proboscis monkey is a pot-bellied, red-faced primate with a bulbous, hanging nose who lives in rain forests near rivers and mangroves on the island of Borneo. Proboscis monkeys eat mostly leaves and some seeds; their leaf-eating ways contribute to this primate's odd, pot-bellied appearance. Plant cell walls are largely cellulose, and therefore indigestible for most mammals. Only three groups of mammals can digest cellulose: kangaroos, ruminants (like cows, goats, deer, elk, and sheep), and colobines (including the proboscis monkey, Asian langurs, and African colobus monkeys). Proboscis monkeys have evolved "huge, complex stomach pouches" that contain "colonies of bacteria" to ferment cellulose into fatty acids that can be absorbed into the bloodstream (Groves 2008, 10). They consume great quantities of leaves, all of which must be broken down by bacteria, which causes these red-faced primates to be "grotesquely pot-bellied for most of the day" (Groves 2008, 10). Proboscis monkeys are endangered because we buy palm oil and tropical hardwoods. Have you noticed how many common products contain palm oil?

Tarsiers can rotate their heads 180 degrees ("Asian Animals"). They are no larger than mice, weighing only four to five ounces. Nonetheless, they can leap as much as ten feet, landing on froglike toes at the tips of boney fingers ("Pygmy Tarsiers" 2009, 8). Tarsiers are named after the sturdy tarsal (ankle) bones that allow these diminutive creatures to jump many times their height. These little wide-eyed hoppers are nocturnal predators, leaping through the twilight in feverish pursuit of cockroaches, crickets, and small reptiles.

"The tarsiers fall somewhere between the lemurs and monkeys on the evolutionary scale" (Ramos). There are several species, all of them found only on Southeast Asian islands. Pygmy tarsiers are now so rare that not a single individual was sighted between 1921 and 2008.

The Javan langur, native to the rain forests and mangrove swamps of Indonesia, lives in small groups, each protected by one adult male. They are blond and black, their faces surrounded by a great mane of beautiful hair. In 2004 twenty-five hundred Javan langurs were poached from the wilds of Indonesia for the pet trade, food, and medicines. Additionally, their habitat is degraded and rapidly disappearing because of human sprawl (Nursahid 2008, 18).

Ruffed lemurs are stark black and white with bright yellow eyes and a ruff of long white fur around their ears and neck. They live in trees and move about on all fours, leaping through the upper canopy of Madagascar's rain forests in search of fruits and filling the air with raucous calls. They are territorial and live in small groups of generally less than half a dozen individuals. Unlike most primates, ruffed lemurs give birth to groups of two or three offspring, but similar to other species, their newborns are helpless, requiring parental protection and tender nurture. Because they cannot carry all of their babies at once, ruffed lemurs are the only primates (other than human beings) who build nests where they can leave their infants while they seek food. These individuals are endangered by both logging and hunting.

Some one thousand lorises are trafficked from Sumatra to Jakarta annually in the illegal Indonesian pet trade ("Vet Describes" 2008, 8). Evidence suggests that six to seven thousand of these petite primates are poached from the wilds of Thailand each year ("Slow Lorises" 2007, 15).

Lorises have little round ears, large round eyes, and a button nose. Because they look so adorable to most humans, uninformed citizens purchase lorises as "pets." However, they make very poor pets: lorises are nocturnal, so they sleep when humans are awake. Also problematic, they do not consume human foods but eat only insects. Nonetheless, they continue to be popular as pets. In May of 2007, forty Thai lorises were kidnapped from the wild and exported to Japan, where they were discovered and confiscated; unfortunately, twelve of these little refugees died before they could be rescued.

How many of us have heard of Bioko or know where it is? Bioko is a small, isolated, volcanic island in the Gulf of Guinea. It is now part of Equatorial Guinea—Africa's only Spanish-speaking country. Bioko's nonhuman primates are unique throughout the world—found nowhere else. They include the Bioko drill, Bioko black colobus, Pennant's red colobus, the red-eared guenon, Bioko crowned guenon, Stampfli's putty-nosed guenon, Preuss's guenon, Bioko pallid needle-clawed galago, Allen's squirrel galago, Prince Demidoff's galago, and Thomas's galago (Weinberg 2007, 12). Although most people have never heard of these magnificent species, they require our attention if we are to avert their extinction.

So many endangered primates—so little media attention. In spite of the media's disinterest, the International Primatological Society (IPS) regularly reports on the most threatened primates and lists them according to country:

- Madagascar: greater bamboo lemur, silky sifaka lemur;
- Africa: roloway monkey, Tana River red colobus, Cross River gorilla;
- Asia: Cat Ba langur, western hoolock gibbon, Sumatran orangutan;
- South America: Peruvian yellow-tailed woolly monkey, brown-headed spider monkey (McGreal 2008, 8).

Considering the extreme danger these primates face and the finality of extinction, it is disconcerting that so few people have heard anything about these species or learned how our activities threaten their existence.

Collectively we are not only ignorant of other primates, but we are also chillingly indifferent: Seventy-five thousand primates are currently exploited for U.S. testing and research (Thirlway 2009b, 14–15). Citizens know that white-clad lab workers experiment on other primates but do not insist on a change of policy. Scientists treat these primates like petri dishes—recording their emotional distress and physiological terror as if these reactions were weather fluctuations on a barometer. The international trade in primates flourishes largely because we continue to exploit these individuals for science. In July of 2007, 950 crab-eating macaques—captured from the wilds of Malaysia—were reclaimed from smugglers. By the time they were seized, more than 100 had suffocated, and other starving captives had consumed the dead. It is likely that these macaques were headed for China, ultimately on their way to research laboratories in Western countries ("Hundreds" 2007, 12).

Scientific exploitation and consumer indifference are evident in the shocking 28,091 primates brought into the United States in 2008; more than 99 percent of them "were for use in scientific research or pharmaceutical testing." The primary importers were Covance Research Products, Inc. (11,360) and Charles River Laboratories, Brf. (7,712) (Thirlway 2009b, 14–15). Many were imported to become the victims of gruesome biowarfare experiments. These victims of "science" "will live lives of pain and suffering and die young" ("Chinese Monkey" 2008, 11). The crab-eating macaque is the most common primate in science labs; the United States imported 26,512 of these individuals in 2008 with China, which supplied 18,087, as the main provider. Other large sources included

Mauritius (4,502), Cambodia (1,920), and Vietnam (1,800) (Thirlway 2009b, 14–15).

What is particularly alarming about these figures is that crab-eating macaques are not native to China. China imports macaques from nearby countries, perhaps Vietnam and Cambodia, or even Indonesia, using falsified documents claiming that these primates are "captive bred." Such trafficking causes "catastrophic declines in the populations of wild monkeys in their native lands" ("Chinese Report" 2009, 12–13). If we are going to protect endangered species, we must ban the use of primates in laboratories.

Zoos are also part of the problem of diminishing wild-primate populations. In 2008 zoos bought less than 1 percent of the imported primates. But in May of 2006, six United States zoos imported thirty-three feral primates, paying more than twelve thousand dollars for a single individual. These zoos (San Diego, Wildlife World in Arizona, Denver, Lowry Park in Florida, and Houston and San Antonio in Texas) admitted that these monkeys were caught in the wild but justified their actions by saying that these individuals were victims of the bushmeat trade with nowhere else to go. Whatever excuses these zoos offered, buying primates supports consumer-based trade, which encourages poachers and dealers to kidnap primates from their wild homes and smuggle them abroad ("U.S. Zoos" 2006, 20–21). Zoos must not buy primates.

U.S. laws encourage smugglers to deal in primates. Not only are imports permitted for research and zoos, but primates can be legally sold as pets within the country. Consequently, primates are bred in the U.S. and can be bought and sold. Once a gibbon or loris has been successfully smuggled into the country, this endangered species can also be sold as a pet without fear of reprisal. According to Born Free USA, only twenty-one states have thus far banned pet trade in primates while twelve others merely regulate possession; seventeen states have no laws whatsoever concerning primates ("Pet Chimp" 2009, 21). If we want to protect these endangered species abroad, the United States must ban both domestic primate breeding and trade. When there is no internal trade in primates, no gibbons, lorises, macaques, or chimpanzees will be captured from their lush homes to be smuggled across U.S. borders.

If we are to pull nonhuman primate species back from the brink of extinction, we must become informed. This is not easy when our government lacks transparency. It is unclear how many primates are currently being exploited by

U.S. scientists; estimates stand at around seventy to seventy-five thousand individuals annually ("Future" 2008, 14). It is also easy for citizens to be fooled into complacency when meager improvements are promised. For example, the U.S. government recently committed to the lifetime care of "surplus" chimpanzees who have been exploited for federally funded research in the 2000 Chimpanzee Health Improvement, Maintenance, and Protection Act. This sounds promising until we learn that chimps comprise only 2 percent of the primates exploited for research in the U.S. ("Chimpanzee Research"). In 2008 the U.S. House of Representatives introduced legislation to "prohibit laboratory testing on all apes." But again, very few apes are forced into research (Crair 2008, 7). What of the 26,512 crab-eating macaques imported for science in just one year?

In contrast to U.S. trends, scientific exploitation of primates peaked in the United Kingdom in 2006 at 7,392; the U.K. used just 1,244 primates in 2008, largely for pharmaceutical safety tests. While even one individual is too many, this number is not nearly as ugly as United States figures.

Similarly the parliament of the European Union adopted a declaration in 2007, drafted by Animal Defenders International, calling for an end to the use of apes and wild-caught monkeys in laboratories ("European Parliament" 2007, 22). This document was signed by 433 members of the European Parliament (MEPs) and includes "a timetable for the phasing out of the use of all primates" (Thirlway 2009a, 24). Today roughly one thousand (10 percent) of the primates exploited in labs in the European Union are caught in the wild ("Promising Proposal" 2009, 9). Unfortunately, European Union laws do not protect primates who have been bred in captivity from exploitation in labs.

While the European Union is well ahead of the United States, the exploitative nature of humanity remains both a concern and a disappointment. What led us to believe that it is morally acceptable to exploit individuals from other species for scientific experimentation? We must revisit our assumptions. Is it morally acceptable to exploit other humans in harmful ways in the hope of learning something that may be useful? While it may be true that many humans would not hesitate to experiment on nonhumans in the hope of saving a loved one, it is equally true that many would just as eagerly experiment on humans to sustain this same hope. Desperation does not shape ethics; desperation is just one of many good reasons why we must establish and hold firm ethical principles.

The fact that we are too often willing to exploit and abuse others to benefit ourselves is not a legitimate moral argument for animal experimentation. On the contrary, because we tend to be selfish, morality—and resultant laws—ought to protect the vulnerable against those who are more powerful and might choose to exploit individuals for their own ends. Might does not make right; self-interest—even in desperation—does not justify exploiting others, whether they come from a different race (Jews in Germany, Native Americans in the United States, Tibetans in China) or a different species.

Other primates seem to understand this moral truth. A study published in 1964 in *The American Journal of Psychiatry,* conducted on macaques at Northwestern University Medical School, was designed to assess altruistic behavior. Scientists created an experiment where macaques could eat only "if they were willing to pull a chain and electrically shock an unrelated macaque whose agony was in plain view through a one-way mirror" (Sagan and Druyan 1992, 117). Using a three-second "5 ma high-frequency shock," the macaques quickly learned that the chain providing food also harmed another individual. One macaque, after witnessing the effects of the shock on another, refrained from "manipulating *either* chain for 5 days and another for 12 days" (Masserman, Wechkin, and Terris 1964, 584; italics in original). Many other macaques refused to pull the chain; in effect they refused to eat: "in one experiment only 13% would do so, 87% preferred to go hungry. One macaque went without food for nearly two weeks rather than hurt its fellow" (Sagan and Druyan 1992, 117).

In this experiment, scientists "demonstrated that most rhesus monkeys refrained from operating a device for securing food if this caused another monkey to suffer an electric shock" (Masserman, Wechkin, and Terris 1964, 584). The scientists concluded, "A majority of rhesus monkeys will consistently suffer hunger rather than secure food at the expense of electroshock to a conspecific" (585). It is ironic that humans shamelessly traumatized these nonhumans to see if they have a sense of altruism and morality; in the final analysis, it is *our* morality—our capacity for compassion and empathy—that is noticeably lacking.

In a study conducted with human subjects, designed by Stanley Milgram at Yale in 1963 and published in the *Journal of Abnormal and Social Psychology,* an authority figure (a man in a white coat) instructed participants to act in ways that were both cruel and potentially dangerous to another human being. The goal was to discover "how much pain an ordinary citizen would inflict on another person" simply because someone in a white coat ordered him or her to do so ("Milgram Experiment").

The subjects were told that the experiment explored learning ability. They were instructed to purposefully and directly shock another individual—the "learner"—whenever he or she made an error. Participants could not see the learner, but they heard the response to the shocks that they *thought* they administered. (Unlike the macaques, the humans here were not actually shocked. Their dramatic responses were prerecorded by an actor specifically for this experiment.) "After a number of voltage level increases, the actor banged on the wall that separated him from the subject." After banging several times "and complaining about his heart condition," the prerecorded learner ceased making any responses ("Milgram Experiment").

In spite of these chilling recordings, most subjects continued to "shock" the learner—with encouragement from the man in the white coat—even after the learner fell silent. In fact, 65 percent of the subjects administered the maximum shock—a potentially deadly 450 volts. While at some point every subject questioned what he or she was doing, only one "steadfastly refused to administer shocks *before* the 300-volt level." Perhaps most importantly, not even one subject "insisted that the experiment itself be terminated" ("Milgram Experiment").

Another experiment using human subjects—this time at Stanford under Philip Zimbardo in 1971—provided twenty-four white middle-class males with roles in an artificial prison community. The intent was to explore the effects of institutional roles on individuals over a two-week period. In the basement of one of Stanford's buildings, Zimbardo created an artificial prison, where volunteers randomly became either prisoners or guards—and they took to their roles famously. Guards were soon engaged in sadistic behavior, humiliating, harassing, and abusing inmates, who were often required to live in filth. They readily administered physical punishment, even forcing prisoners "to go nude as a method of degradation, and some were subjected to sexual humiliation, including simulated homosexual sex" ("Stanford Prison").

Zimbardo himself, assuming the role of prison superintendent, was unmoved in the face of increasing cruelty and sadism. Roughly fifty people visited the prison while it was in session, but not one suggested—let alone insisted—that the experiment be terminated. Then Zimbardo's girlfriend, Christina

Maslach, came to visit, and she "objected to the appalling conditions," causing Zimbardo to conclude the experiment after just six days.

While both of these experiments were designed to study the effects of outside forces on the human moral compass (an authority in a white coat or a role in one's institution), comparing the Yale and Stanford studies with the macaque experiment is humbling. In contrast with human primates, the macaque primates showed "a saintly willingness to make sacrifices in order to save others" (Sagan and Druyan 1992, 117). In the words of Sagan and Druyan,

> macaques—who have never gone to Sunday school, never heard of the Ten Commandments, never squirmed through a single junior high school civics lesson—seem exemplary in their moral grounding and their courageous resistance to evil. Among the macaques, at least in this case, heroism is the norm. If the circumstances were reversed, and captive humans were offered the same deal by macaque scientists, would we do as well? In human history there are precious few whose memory we revere because they knowingly sacrificed themselves for others. For each of them, there are multitudes who did nothing. (117–18)

With such unnerving evidence from our labs, one wonders how humans can possibly believe that those placed in positions of power and authority can be trusted to know when animal experiments ought to be terminated for moral reasons. Such morally questionable experiments help us to ponder what it means to be one primate among many. Increasingly we come to understand that any comparison between human and nonhuman primates does not show humans in a complimentary light. And any difference between the two is merely one of degree with regard to reason, altruism, or language, for example. Differences are of degree, not kind, and humans are not the pinnacle of evolution that we have too often imagined ourselves to be.

For this reason, philosopher Tom Regan extends rights to nonhuman primates (and many other species). He defines "persons" as individuals who have an experiential welfare—they fare better or worse depending on their circumstances. He notes that we are persons because "each of us is equally a somebody, not a something" (Regan 2003, 81). What happens to persons matters—"whether to our bodies, or our freedom, or our lives themselves—[it]

matters to us because it makes a difference to the quality and duration of our lives, as experienced" (Regan 2005, 50). Persons—somebodies—are affected by circumstances. Just as our welfare matters to us, the welfare of a stray poodle, factory-farmed hog, or starving primate matters to each of them. For example, macaques are negatively affected by the experiences of an electric shock and starvation. Pygmy lemurs and proboscis monkeys also have experiences; their lives fare better or worse based on what happens to them—whether they are enslaved as pets or robbed of their homes by diminishing habitat. Each of these primates is a person—a somebody, not a something.

Because circumstances and experiences affect the welfare of persons, what we do to them has moral significance. If I enslave you or a gibbon for entertainment, I have behaved selfishly and cruelly because I have diminished your/his or her life to enhance my own. If I kidnap and sell you or a chimpanzee, I have behaved selfishly and cruelly because I have disrupted your/his or her community, family, and life for my economic benefit. If I use your body or a macaque's body in the hope of saving myself (or those like me), I have behaved selfishly and cruelly by sacrificing your/his or her happiness, autonomy, and perhaps life to enhance my happiness or lifespan or those of my species or race. It is selfish to exploit others for our gain, whether they are other mammals, other primates, or other human beings. No moral code of conduct encourages selfishness, cruelty, or exploitation.

For these reasons, persons—those affected by their circumstances—have inherent value according to Regan, and those with inherent value have rights "whether or not anybody else cares" (Regan 2003, 82). He defends the rights of nonhumans for the same reason that he defends human rights: you and I, Hereford cattle and white leghorn chickens, and vervet monkeys exploited by the U.S. Department of Defense are all persons ("War" 2009, 6), individuals affected by our circumstances and experiences. We are somebodies—not somethings—individuals who experience our lives as painful or peaceful, chaotic or quiet.

The line that much of the Western world has struggled to maintain between human beings and other animals has always been an artificial construct. We have created and maintained this line to hold ourselves apart and above other animals and to justify exploiting them for our ends: food, clothing, or experimentation. If we are going to save endangered primates, we must first recognize that they are individuals much like human beings, who prefer to be free to live their lives independent of exploitation.

Primate People is about nonhuman primates; this book is also about human beings. Other primates are people, and people are primates. Each is a unique individual; each is legitimately the subject of moral concern.

If we are to honor their rights as individuals and save nonhuman primates from extinction, we must first be aware that they exist, that they are in danger, and that their survival depends on our individual choices. We must understand how our actions threaten primates—and we must *alter our behavior*.

Primates are many and wondrous, yet few and endangered. Everywhere they live, they have been crowded out of diminishing forests, hunted for food or medicine, captured for the lucrative pet/tourist trade, and either kidnapped or bred for science. As a result, every primate species on the planet—aside from human beings—is either endangered or threatened.

Our efforts to protect primates will be much more effective if we dismantle the artificial line that we have created between ourselves and other animals, if we recognize that the lives of nonhumans matter not just to us—not just in light of our selfish interest in diversity—but to them. We are therefore morally required to stop systematically exploiting others, whether they are chimpanzees or pygmy lemurs, chickens or chinchillas. Because nonhuman animals are also persons who fare better or worse depending on the way we treat them, we must begin to give them the respect and dignity that persons deserve.

Authors and Essays

Many authors represented in *Primate People* are activists who are unaccustomed to writing essays for anthologies. They graciously squeezed in a little writing time between tending woolly monkeys or just before departing for a primate foray into the Indonesian jungle. Other authors know English only as a second (or third) language. Consequently, they frequently submitted a rough draft and left polishing to me while they flew across continents to plead on behalf of macaques or rushed to cradle bushmeat orphans. With the help of e-mail and by working together, we turned their understanding, experience, and knowledge into the chapters of this anthology.

Those who submitted these essays work for a variety of primates around the world in an assortment of ways. They are undercover agents, scholars, and researchers; they work in sanctuaries or with grassroots organizations to fight vivisection or lobby for laws that protect primate habitat. Authors in this volume live and/or work in Malaysia, Spain, Thailand, England, Wales, South Africa, Colombia, Denmark, the United States, and Indonesia. They work with baboons, woolly monkeys, capuchins, gibbons, gorillas, macaques, owl monkeys, lemurs, lorises, De Brazza's monkeys, chimpanzees, and spider monkeys.

These authors bring primates to life as individuals and communities, for instance, the baboon who was caught in a snare yet approached a human friend, "holding his arm out" in hopes of help, and a troop of chimpanzees at a sanctuary who exemplify the hopes and hurts of their species in their interactions with their caregiver. Their stories introduce readers to the antics and pleasures, tendencies and idiosyncrasies, sufferings and fears of nonhuman primates, and they explain how humans endanger and harm these close cousins of ours. This last element—how humans endanger and harm primates—is of utmost importance because these stories carry home to readers the effects of our choices. The authors in this volume help us to understand what we can and must do to protect these vulnerable individuals. The essays in this anthology are divided into three sections.

Part I: Foundations

The first section introduces some of the key problems threatening and devastating nonhuman primates, such as the entertainment and pet industries, logging and the bushmeat trade, and habitat destruction caused by our dietary choices. These essays introduce individuals from a range of primate species around the world and explore issues that reemerge in later sections. They highlight fundamental problems facing primates, problems that stem from human ignorance, greed, and indifference. These authors share what they have learned from working with and for primates. We begin with a short essay that introduces primate basics, written by field primatologist Linda Wolfe. She charts the evolutionary history of primates and their social systems, allowing us to reflect on where we fit into the primate family—humans are "territorial pairs" among primate species. She also prepares us for the next essay by introducing CITES, the Convention on International Trade in Endangered Species of Wild Fauna and Flora.

Danish biologist Birgith Sloth focuses on the illegal trade in primates. Sloth is an expert on CITES, a fairly recent international wildlife treaty that attempts to restrict trade in endangered plants and animals. CITES lists all nonhuman primates either under Appendix I or II ("CITES" 2008, 8), and she explains why: thirty-seven primates are now critically endangered, seventy-eight species are at high risk of extinction in the wild, and we lack sufficient information on fifty-six other species to be able to assess their level of danger. Nonetheless, it is legal to trade primates in the international market in the name of science, so they are smuggled across borders for the pet trade, further threatening danger-ously depleted species. Sloth describes the contents of this important document, its implementation, the challenge of enforcement, and how CITES relates specif-ically to the protection and preservation of nonhuman primates.

Phaik Kee Lim lives in Malaysia, a nation where primates are indigenous and some citizens view these relatives of ours as a nuisance. She explains how and why local officials shoot macaques and describes the unfortunate and entirely unnec-essary demise of a dusky leaf monkey, who clutched her infant desperately even in death. She also provides examples of the mistreatment of nonhumans in Ma-laysian circuses, resorts, and zoos and highlights the ongoing illegal trafficking of endangered wildlife, including orangutans and gorillas, across international bor-ders despite the establishment of CITES. Lim works with a Malaysian conser-vation organization to "correct problems affecting humans, nonhumans, and the environment." With a pen that demands change, she writes letters and editorials to warn people away from investing in exotic pets and paying entrance fees to any form of entertainment that supports animal exploitation. With her words, she hopes to stir those in power to act on behalf of orangutans forced to perform frivolous tricks, gorillas kidnapped from their homelands, and macaques deval-ued as pests. Reflecting on twenty-five years of activism, Lim laments, "Nonhu-mans are always at the mercy of humans."

Noga and Sam Shanee bring primate issues closer to home for readers in Eu-rope and North America by linking our choice of foods to tropical deforestation in Brazil, thereby implicating our diet in the plight of one of the most endan-gered primates on the planet: yellow-tailed woolly monkeys. While working on a short research project in Peru, the Shanees learned that yellow-tailed woolly monkeys have been pushed to the precipice of extinction. Charmed by the bold curiosity of these highly endangered primates, they established a conservation organization in La Esperanza, Peru, where they work with locals—especially churches—to turn the tide for these cheeky inhabitants of the Brazilian rain forest.

After the Shanees set up shop in South America, they quickly discovered that big government and big industry have united in their efforts to reap the last dollar from Peru, even at the cost of shoving these obstreperous primates into the abyss of extinction. The Shanees demonstrate the importance of gaining a local perspective and the need for on-site action. Their essay helps readers think about the way our daily choices affect primates on distant continents and recognize that individual initiative and direct action are essential if we are to save dwindling tropical forests and their endangered residents.

Few people, if any, have been as effective in helping raise awareness and bring change for primates as Shirley McGreal. Thirty-five years ago, McGreal met nonhuman primates for the first time, and no one could have guessed how a few tiny stump-tailed macaques, peeking piteously out of a cage at the Bangkok International Airport, would change her life—and the lives of primates around the world. Their frightened eyes stirred curiosity and compassion in McGreal and refocused her youthful energy. With the help of a like-minded friend, McGreal soon discovered a ring of smugglers who were stealing baby gibbons from Thailand's jungles to send them to Singapore, where they were redirected across the ocean to California laboratories. She and her friend exposed and effectively destroyed the Singapore Connection, an illegal transport route that landed Thailand's primates in U.S. labs. Ultimately, the urgency of those caged eyes led McGreal to establish the International Primate Protection League (IPPL), and she has continued to work courageously and effectively on behalf of primates ever since.

Part II: Research

The next set of essays exposes the exploitation of nonhuman primates in research facilities. This section begins with two essays written by animal activists, including an undercover agent who spent time in a primate lab. The third essay exposes a notorious Colombian scientist—Manuel Elkin Patarroyo—and offers an alternative research model. The last two essays in this section continue this general theme, moving from conventional scientific exploitation of primates to more compassionate and moral—not to mention sustainable and effective—scientific models and methods.

As an animal liberationist and dedicated activist, Michael Budkie has meticulously reviewed "tens of thousands" of "inspection reports, research protocols, and health-care records for dogs, cats, goats, and primates" in U.S. research labs—cryptic records describing the physical actions and reactions of individuals sold into science for medical purposes. Budkie provides samples from these disconcerting documents: "primate #312A: 'still overdosing on current drug dosage, ataxic, hypersalivating, disoriented.'" His revulsion at what he has learned about animal research is palpable. Having examined and lobbied against animal experimentation for twenty years, Budkie writes, "I can never forget that each one of the thousands of pieces of paper that I have read—documents detailing the horror of animal experimentation—actually describes the life of an individual."

As an undercover investigator on a two-year assignment for In Defense of Animals, Matt Rossell worked as a lab technician at Oregon Health and Science University's Oregon Regional Primate Research Center (now the Oregon National Primate Research Center), a facility holding twenty-five hundred primates. Ostensibly he was hired to provide enrichment for caged macaques with the stated goal of easing abnormal behavior—actions that are both common and normal for lab primates. Rossell notes that, in truth, he was hired merely to "create a paper trail to meet the hollow requirements of the Animal Welfare Act." He remembers one victim in particular: macaque number 16162, who suffered day by day—depressed, lonely, stressed, and bored—until she fell ill. His essay offers a chilling view inside a primate research facility and reminds us of the innumerable individuals who suffer and die in these stainless-steel facilities.

Juan Pablo Perea-Rodriguez of Colombia explores the motivation and practices of a specific scientist who continues to exploit primates: Manuel Elkin Patarroyo. Like most researchers, Patarroyo hoped to bring great knowledge and innovation to humanity—while gaining wealth and notoriety. Indeed, he found fame and riches, but he provided nothing of value in exchange. Instead, Patarroyo has damaged—and continues to harm—the fragile South American ecosystem and many individual primates. Perea-Rodriguez also describes his student internship at DuMond Conservancy for Primates and Tropical Forests, a facility in Miami, Florida, that uses noninvasive research to educate the public about primates and their habitat. At Dumond his life was unexpectedly transformed when he learned and practiced a scientific approach where humans respect other primates and consider science only a byproduct of their primary role as caretakers and protectors of individuals and habitats.

Anthropologist and primatologist Linda Wolfe transports us to the fresh air and freedom of field research. Watching primates in the field has taught Wolfe not only about macaques but also about humanity—including herself—and animals more generally. Reflecting on her fieldwork, Wolfe ponders human exploitation of nonhumans: "How much torment may we inflict, and for how long, on one helpless individual in a laboratory?" Her writing courageously travels along a razor's edge between the old model of dispassionate, selfish science and her growing understanding of macaques—and all animals—as unique individuals worthy of respect and protection.

Ethologist and primate behaviorist Debra Durham carries us with her into a primate lab, then onward and outward to field research, and finally to the life of an animal liberationist. When she graduated from college, Durham was excited about working in a laboratory. Despite her enthusiasm, she quickly realized that all was not right with the little motherless macaques and baboons who clung desperately to surrogate mothers: PVC pipes covered with cloth in otherwise barren cages. Nonetheless, she fed, watered, measured, weighed, and cuddled these unhappy babies until her conscience forced her out of the lab and back to college. As a graduate student, she traveled to Madagascar to study lemurs in the field— free individuals in their own habitat. The more Durham learned, the more she understood the injustice of primate research labs. Determined to bring change, she took a job with PETA (People for the Ethical Treatment of Animals) and was placed on a case involving government research with a group of rhesus macaques, which she followed for nearly two years. In the process, she came to know Patrick and Brigit, two monkeys exploited and destroyed in the name of science:

> Both were forced subjects of invasive brain experiments. Both had holes cut in their skulls that were fitted with metal guide tubes. The tubes held electrodes that were inserted into their brains during experiments to study brain activity. Patrick and Brigit also had bolts drilled into their skulls and a metal coil implanted in one eye. Four or five days each week, these individuals were strapped into a chair in full-body restraint with their heads bolted in place.

Repulsed by what she learned in her work for PETA, Durham took a job with the Physicians Committee for Responsible Medicine (PCRM), working on behalf of individuals like Patrick and Brigit.

Part III: Sanctuaries

The final section of this anthology focuses on sanctuaries where activists lobby for change, educate communities, and care for displaced nonhuman primates. The first four essays are written by or about sanctuary founders and their work. These essays explain how founders arrived at such a demanding and rewarding career, describe the primates they work with (including profiles of cherished residents), and reveal what these energetic and extraordinary people have learned along the way. These writers exemplify the enduring concern and exceptional dedication that lie behind a career in animal advocacy, in this case, founding and maintaining a sanctuary. The next five essays are written by people who have taken temporary or permanent positions at sanctuaries, whether as volunteers or skilled staff. These writers describe what employees and volunteers do at sanctuaries in places such as Southeast Asia and South America and what readers can expect if they choose to do similar work.

Barbara Cox begins a series of essays that focus on individual primate sanctuaries established as permanent residences and their founders. She writes about a Florida sanctuary for Central and South American primates and walks us through the misguided moments that carry primates from freedom and health into the unwitting arms of untrained humans in ill-suited homes. Unbeknownst to Kari Bagnall, Jungle Friends Primate Sanctuary began when her boyfriend bought a baby capuchin, whom she named Samantha. He quickly grew tired of the little troublemaker, only to find that his girlfriend—when forced to choose between him and the difficult, diminutive capuchin—sent him packing. Samantha proceeded to destroy Bagnall's house and even sank her sharp teeth into Bagnall's visitors.

Still not grasping the core problem, Bagnall purchased Charlotte, a "sister" for Samantha, and the two youthful capuchins quickly demolished everything from water fountains and pools to private bedrooms. Eventually Bagnall got the picture: capuchins are neither pets nor human children. She also discovered that she was not the first person to be duped by the pet-trade industry: there are many more pet capuchins than sanctuary openings. Having unwittingly contributed to the problem, Bagnall made a remarkable commitment: she accepted long-term responsibility and created Jungle Friends Primate Sanctuary.

From the loss of her own newborn son to the moment she first sang with a gibbon, Deborah Misotti healed alongside damaged and exploited primates. Misotti's sanctuary dreams were born at a facility that exploited primates for entertainment, and she explains the way a proper sanctuary differs from such a capitalistic enterprise. She defines a sanctuary as "a place of refuge or asylum," where residents are not owned, harassed, or exploited. Misotti created, and now manages, a Florida sanctuary that provides lifelong care for primate victims of the trade in exotic pets, research laboratories, and/or breeding facilities. She ponders the arrival of Chi Chi, a stunning black gibbon who was uprooted and transported from one capitalist venture to another: exploited for her reproductive ability, repeatedly impregnated, perpetually pregnant, yet never allowed to be a mother. Misotti is painfully aware that this primate has been denied the simple and seemingly inalienable right to have the life of "a gibbon without the constraints of human greed, ownership, and intrusion."

The next author, Rita Miljo, founded CARE (Centre for Animal Rehabilitation and Education) in South Africa. In spite of her ongoing efforts to increase local awareness, baboons are considered vermin in South Africa and can be exterminated with guns, traps, or poison ("IPPL Members" 2006, 5). She describes a nearby colony of wild baboons and the devastating effect of snares on these vulnerable and much maligned individuals. Miljo tells readers about two baboons who came to her from research labs, another who spent most of her life "welded into a forty-five-gallon drum," and a little orphaned baboon who, though released into the wild, never forgot the comforts of CARE—or her human caretaker. She takes us with her through some of the inevitable, but agonizing, life-and-death decisions that she must make on behalf of individuals who have come under her care.

We close our exploration of sanctuary founders with a visit to Indonesia under the guidance of Spanish sanctuary veterinarian Karmele Llano Sanchez. Sanchez worked with primates in Venezuela and Holland before cofounding a primate center in Indonesia, where she specializes in rehabilitation and release of macaques and lorises into the wild. She describes the capitalistic enterprises endangering these two primate species: local and international markets for exotic pets, research, and delicacy dishes and the local development of palm plantations that claim vital habitat, where owners and operators kill local primates. Sanchez also explains what happens to lorises in the Indonesian pet industry and describes her veterinary efforts on behalf of these unfortunate victims of capitalism and consumer ignorance. She details the costs inherent in running a sanctuary and describes some of the individual residents she has dealt with as a sanctuary veterinarian, including an orangutan with a fractured skull and a newborn gibbon with a raging fever. Sanchez expresses strong compassion for her patients and palpable frustration with the human insensitivity and greed that underlie the painful problems she works to cure.

Most readers are not in a position to found a sanctuary, but many can do what Fiona Mikowski has done. As a college student, she volunteered at the Gibbon Rehabilitation Project (GRP) in Thailand. She describes what she experienced as a volunteer and visitor in Thailand: "Though living in extreme humidity in the midst of an unforgiving rainy season thick with mosquitoes, cockroaches, frogs, and snakes, I loved my time at the GRP," she says. Mikowski vividly portrays the daily chores of rehabilitation and the thrill of watching sanctuary residents move from life in a cage back to their rightful home in the jungles of Thailand. She also introduces us to three permanent residents at GRP—victims of the profit-driven trade in primates as pets: Tam, Bo, and Joy. In the process, Mikowski exposes the illegal transport of gibbons from their lush jungle homes to dingy basement cages, where these athletic little apes often languish for years.

Young people like Fiona Mikowski sometimes build on their initial experience in sanctuaries to become primate experts in their own right. There is tremendous need for individuals who specialize in nonhuman-primate care and rehabilitation like veterinarian Karmele Llano Sanchez, ethologist Debra Durham, and student intern Juan Pablo Perea-Rodriguez. The next three essays are written by these trained and experienced individuals, who provide much-needed temporary help at three very different sanctuaries, one in the United States, one in Peru, and one in Ireland.

Paula Muellner was giddy with excitement when she drove across the U.S. to take up her new post at a chimpanzee sanctuary in Oregon, but when she arrived, the chimpanzees spit water at her and flung "meticulously prepared" smoothies back in her face. Luckily Muellner understood that these chimpanzees had been exploited, abused, neglected, and abandoned repeatedly by humans; it would take time to earn their trust. After months of food throwing, the chimpanzees slowly began to invite her to play and groom. Herbie eventually took a shine to Muellner…little did he know that she was only a temporary caretaker. Muellner reflects on the depth of emotions expressed by Herbie and the other sanctuary residents and the moral concerns and complications of taking a two-year position working with individuals who form deep and lasting bonds—and have no way of knowing that you are only part of their community for a short time. On leaving, she realizes, "I had broken the circle of trust that I had worked so hard to build."

Experienced primate caretaker Keri Cairns flew from the United Kingdom to Ikamaperu, a monkey sanctuary in Peru, to tend residents and build an enclosure while founders and primary caretakers Carlos and Helene Palomino were away for six weeks. With an eye to colonies of termites and intense earthquakes, Cairns designed and built a sturdy primate enclosure on the edge of the Peruvian rain forest, but he also got to know the local residents. He befriended a large adult male woolly monkey, Apu, who had all the teeth and strength needed to kill a human being. Yet Cairns slowly gained Apu's trust, and in turn, learned to trust Apu, even sharing playful moments with this powerful individual from another species.

Cairns explains the importance of sanctuaries and reminds readers that our choices affect primates. He writes that woolly monkey habitat is "cleared for agricultural purposes such as grazing beef cattle or growing soya to feed these cattle for market. I became a vegetarian twenty-five years ago after reading about the effect that this production of cheap beef, mainly for Western fast-food outlets, was having on the Amazon rain forest." A zoologist skilled in the care of primates, Cairns reminds readers that, whether they are teachers, engineers, architects, writers—or just plain handy with a few tools—their skills are greatly needed for animal advocacy.

Helen Thirlway, now head of IPPL in England, writes about a particular monkey, Singe, whom she met while working at a sanctuary in Ireland. Singe was a pudgy, disillusioned De Brazza monkey who had spent twenty long years lounging on overstuffed furniture and eating junk food, enslaved as a pet. As with most exotic pets, Singe's humans ultimately sought to be rid of this living acquisition, in this case turning her over to a primate sanctuary. De Brazzas normally live in central Africa—there are no De Brazza sanctuaries in Europe—so

Singe lived "alone, with no companions of her own kind." Thirlway was given the task of taking care of this dejected resident, trying to bring a bit of joy into Singe's limited and lonely life. To do so, she had to gain Singe's trust. Her essay describes both the challenges that face sanctuary caretakers and the problems that result when humans purchase nonhumans as exotic pets.

Licensed professional civil engineer and Indian American writer and activist Sangamithra Iyer writes, "I was interested in catastrophes—and in preventing them." Iyer communicated in sign language with Washoe and watched research chimps neurotically twist about in their "enriched" cages, she tended chimpanzees orphaned by the bushmeat trade in Cameroon, and she met highly endangered mountain gorillas in Rwanda. While in Rwanda, Iyer also visited a memorial church that houses skulls and other bones of genocide victims, and she talked with Rwandans who visibly wear the scars of unmitigated human violence. As one who has worked cleaning up hazardous wastes, she understands that it takes billions of years to reclaim poisoned soils. Reflecting on her time with orphaned chimpanzees and battle-scarred humans, Iyer ponders the nature of the human primate—our tendency toward violence, our ability to move forward after losses that seem impossible to bear, and the legacy of violence:

> I am not a primatologist, nor a psychologist, but I have seen a chimpanzee spin her head in figure eights, and I have felt little orphaned fingers clinging to my shirt. I have seen skulls lined up on a shelf in a wooden church in Ntarama, and I have witnessed the way we deal with hazardous wastes in the United States. I know our human hands are soiled, and I wonder how we can clean up our messes.

In this highly reflective piece, likely speaking for most of the authors in this book—and certainly speaking for the editor—Iyer writes, "I wanted to know that new lives were possible."

Primate People

Part I

Foundations

1

Primate Basics

Linda D. Wolfe

Primates consist of human beings, apes, monkeys, and prosimians. Prosimians are lemurs and their relatives (galagos and lorises; see charts). Chimpanzees, who are classified as apes, are our closest relatives. They are considered genetically closer to human beings than to any other primate.

Primate fur is usually various shades of brown, although some species, such as gorillas, have black hair, and some, such as the African arboreal monkeys generally known as cercopithecines, have colorful hair around their faces.

True primates first appeared in the Eocene epoch, fifty-five to thirty-five million years ago, during a period of global warming and forest expansion. Early primate fossils have been found in what are now North America and Europe. Many of the Eocene primates resembled the lemurs that live on the island of Madagascar today. Not all early primates left descendants, of course, but those who did are the ancestors of today's primates.

Fossils from the Eocene and the following Oligocene epoch reveal the evolution of basic primate traits. These traits include the following:

- arboreal and diurnal lifestyle (living in trees and remaining active during the day);
- stereoscopic vision (eyes facing forward, providing overlapping fields of vision that offer depth of perception) and eyes encased in protective bone;

- acute hearing;

- eye-to-hand coordination;

- color vision;

- grasping hands and feet;

- nails and small ridges (such as human fingerprints) on the soles of the feet, palms of the hands, and fingers and toes;

- enlarged brain to body ratio;

- social grouping based on intimate bonds;

- delayed maturation and increased longevity;

- and the presence of one or two offspring combined with intensive maternal care.

The first primates were very small, weighing around two hundred grams (seven ounces). Today primates range in size from under two hundred grams, such as the smallest lemurs of Madagascar, to gorillas who weigh more than 175 kilograms (386 pounds).

The original locomotion of Eocene primates was probably a form of movement called vertical clinging and leaping, which was well suited for the long legs and short arms of many lemurs and other prosimians. From this basic pattern, many different types of primate movement evolved, including the brachiation of gibbons, knuckle walking of chimpanzees, and upright walk of *Homo sapiens*.

Beginning about thirty-five million years ago, primate evolution shifted from North America and Europe to Africa, Asia, and Southeast Asia (China and Burma). Today about 250 to 300 species of nonhuman primates live near the equator in tropical forests in Central and South America and forested sections of Africa and Asia, such as India, China, Indonesia, and Malaysia.

Over the last fifty-five million years of primate evolution, the Earth's climate has undergone changes characterized by warmer, wetter, dryer, and cooler periods. Prosimians, monkeys, and apes diversified and adapted accordingly. In fact, based on their genetic heritage, which favored an enlarged brain and basic sociality, primates developed several different types of social systems to meet challenges presented by their environment. Primate social systems are not the same as mating patterns; most primates—both males and females—engage in opportunistic mating and seek an array of different partners. There are a handful of distinct primate social systems:

- Noyau: Females with young have small home ranges that do not overlap with other females. Males have larger ranges that do not overlap with other males but do overlap with one or more female ranges. Examples: nocturnal prosimians and orangutans.

- Territorial pairs: Pairs of males and females occupy an area with their offspring. Core territories of male/female pairs do not overlap with those of other pairs. Example: gibbons.

- Unimale groups: One mature male lives in community with several females of various ages. Examples: hamadryas baboons (and some other baboons) and gorillas.

- Polyandrous groups: One reproducing adult female lives in community with nonreproducing adult females and two or more adult males, all of whom help to care for offspring. Example: tamarins.

- Multimale/multifemale troops: A group of primates, ranging from twenty to more than two hundred individuals, contains multiple adult males, varied females, and their offspring, who stay together throughout the day and night. Examples: macaques and baboons.

- Fission/fusion communities: Adult males, adult females, and their offspring remain in large groups when supported by a large supply of food or fission into small foraging groups when food is scarce. Examples: chimpanzees and spider monkeys.

As primates have evolved, they have diversified and adapted to many different environments, yet all primate species except human beings are now threatened with extinction. Approximately 250 to 300 species of primates remain; all are listed in Appendix I (as endangered) or Appendix II (as threatened) by CITES (the Convention on International Trade in Endangered Species of Wild Fauna and Flora). To cite just a few examples, Appendix I endangered primates include Central and South American howler and spider monkeys, African leaf-eating monkeys, chimpanzees, gorillas, and gibbons in India and Thailand. Primates are endangered and threatened mainly due to habitat destruction, hunting for bushmeat, and the international trade (legal and illegal) in monkeys and apes, both as exotic pets and for biomedical research.

Ecotourism can be both a boon to the survival of wild primates and a further threat. For example, people in primate-habitat countries are more likely to want their wild cousins to survive if they can earn money from tourists who come to view orangutans or gorillas. However, through contact with *Homo sapiens,* these primates contract diseases like measles and polio with devastating effects. And because tourism habituates nonhuman primates to human presence, it also makes these communities more vulnerable to poachers. The future well-being of wild primates depends on our goodwill and foresight.

Comments on Primate Taxonomy

There are two major schemes for the division of the order of Primates. The Anthropoidea/Prosimii (suborders) include the Anthropoidea (monkeys, apes, and people) and the Prosimii (lemurs and related wet-nosed primates, and tarsiers). The Strepsirhini/Haplorhini (suborders) include the Strepsirhini (lemurs and related wet-nosed primates) and the Haplorhini (tarsiers, monkeys, apes, and human beings).

Chart 1. Alternate Taxonomies of the Superfamily Hominoidea (informal names included in brackets)

---

1. Traditional

Superfamily: Hominoidea [hominoid(s)]

    Family: Hylobates—gibbons [hylobate(s)]

    Family: Pongidae—orangutans, gorillas, chimpanzees, bonobos [pongid(s)]

    Family: Hominidae—modern humans [hominid(s)]

2. My Preference

Superfamily: Hominoidea [hominoid(s)]

    Family: Hylobates—gibbons [hylobate(s)]

    Family: Pongidae—orangutans [pongid(s)]

    Family: Panidae—gorillas, chimpanzees, bonobos [panid(s)]

    Family: Hominidae—modern humans [hominid(s)]

3. Currently in Use

Superfamily: Hominoidea [hominoid(s)]

    Family: Hylobates—gibbons [hylobate(s)]

    Family: Hominidae [hominid(s)]

      Subfamily: Ponginae—orangutans [pongine(s)]

      Subfamily: Gorillinae—gorillas [gorilline(s)]

      Subfamily: Homininae—chimpanzees, bonobos, humans [hominine(s)]

        Tribe: Panini—chimpanzees, bonobos [panin(s)]

        Tribe: Hominini—humans [hominin(s)]

          Subtribe: hominine—*Homo sapiens* and ancestral congeners [hominan(s)]

There are two subfamilies of the family Cercopithecidae: Cercopithecinae and Colobinae. The cercopithecines are fruit and insect eaters and include the macaques, baboons, and vervet monkeys plus others. The colobines are leaf eaters, including such primates as colobus monkeys and langurs.

Chart 2. Primate Classification

Phylum: Chordata (vertebrates)

    Class: Mammalia (mammals)

        Order: Primates (lemurs, monkeys, apes, humans)

            Suborder: Prosimii

                Infraorder: Lemuriformes (lemurs, lorises, galagos, sifaka, tarsiers)

                    Family: Cheirogaleidae

                        Sample Genera: *Cheirogaleus, Microcebus, Allocebus, Phaner*

                    Family: Lemuridae

                        Sample Genera: *Lemur, Hapalemur, Eulemur, Varecia*

                    Family: Lepilemuridae

                        Genus: *Lepilemur*

                    Family: Indriidae

                        Genus: *Indri, Propithecus, Avahi*

                Infraorder: Chiromyiformes

                    Family: Daubentoniidae

                        Genus: *Daubentonia*

                Infraorder: Lorisiformes

                    Family: Lorisidae

                        Sample Genera: *Loris, Nycticebus, Arctocebus, Perodicticus*

                    Family: Galagonidae

                        Sample Genera: *Galago, Euoticus, Galagoides, Otolemur*

                Infraorder: Tarsiiformes

                    Family: Tarsiiae

                        Genus: *Tarsius*

             Suborder: Anthropoidea

                Infraorder: Platyrrhini (New World monkeys)

                Superfamily: Ceboidea

                    Family: Cebidae

                        Sample Genera: *Cebus, Saimiri, Callimico, Leontopithecus, Saguinus*

                    Family: Aotidae

                        Genus: *Aotus*

Family: Pitheciidea

Sample Genera: *Pithecia, Chiropotes, Cacajao, Callicebus*

Family: Atelidae

Sample Genera: *Ateles, Lagothrix, Brachyteles, Alouatta*

Infraorder: Catarrhini (Old World primates)

Superfamily: Cercopithecoidea (Old World monkeys)

Family: Cercopithecidae

Sample Genera: *Macaca, Cercopithecus,* Chlorocebus, *Papio, Mandrillus, Colobus, Presbytis, Nasalia, Rhinopithecus*

Superfamily: Hominoidea (apes and humans)

Family: Hylobatidae

Sample Genera: *Hylobates, Symphalangus*

Family: Hominidae

Genus: *Pongo, Pan, Gorilla, Homo*

Chart 3. Primate Classification with Common Names (common names in brackets)

Phylum: Chordata (vertebrates)

Class: Mammalia (mammals)

Order: Primates (lemurs, monkeys, apes, humans)

Suborder: Prosimii

Family: Cheirogaleidae

Sample Genera: *Cheirogaleus* [dwarf lemurs], *Microcebus* [mouse lemurs], *Allocebus* [hairy-eared mouse lemurs] *Phaner* [fork-crowned lemurs]

Family: Lemuridae

Sample Genera: *Lemur* [ring-tailed lemurs], *Hapalemur* [bamboo lemurs], *Eulemur* [lemurs], *Varecia* [ruffed lemurs]

Family: Lepilemuridae

Genus: *Lepilemur* [sportive lemurs]

Family: Indriidae

Genus: *Indri* [indri], *Propithecus* [sifaka], *Avahi* [avahi]

Family: Daubentoniidae

Genus: *Daubentonia* [aye-ayes]

Family: Lorisidae

Sample Genera: *Loris* [lorises], *Nycticebus* [slow lorises], *Arctocebus* [angwantibos], *Perodicticus* [pottos]

Family: Galagonidae

Sample Genera: *Galago* [bush babies], *Euoticus* [needle-clawed bush babies], *Galagoides* [bush babies], *Otolemur* [greater galagos]

Family: Tarsiiae

Genus: *Tarsius* [tarsiers]

Suborder: Anthropoidea

Infraorder: Platyrrhini (New World monkeys)

Family: Cebidae

Sample Genera: *Cebus* [capuchins], *Saimiri* [squirrel monkeys], *Callimico* [Goeld's marmosets], *Leontopithecus* [lion tamarins] *Saguinus* [tamarins]

Family: Aotidae

Genus: *Aotus* [night monkey]

Family: Pitheciidea

Sample Genera: *Pithecia* [saki monkeys], *Chiropotes* [bearded sakis] *Cacajao* [uakaris], *Callicebus* [titis]

Family: Atelidae

Sample Genera: *Ateles* [spider monkeys], *Lagothrix* [woolly monkeys], *Brachyteles* [muriquis], *Alouatta* [howler monkeys]

Infraorder: Catarrhini (Old World primates)

Family: Cercopithecidae

Sample Genera: *Macaca* [macaques], *Cercopithecus* [guenons] *Chlorocebus* [monkeys], *Papio* [baboons], *Mandrillus* [mandrills] *Colobus* [colobus], *Presbytis* [langurs], *Nasalia* [proboscis monkeys], *Rhinopithecus* [snub-nosed monkeys]

Superfamily: Hominoidea (apes and humans)

Family: Hylobatidae

Sample Genera: *Hylobates* [gibbons], *Symphalangus* [siamangs]

Family: Hominidae

Genus: *Pongo* [orangutans], *Pan* [chimpanzees, bonobos], *Gorilla* [gorillas], *Homo* [humans]

2

International Primate Conservation

*The Convention on International Trade in Endangered Species
of Wild Fauna and Flora (CITES)*

Birgith Sloth

Introduction

In the second half of the last century, wildlife trade increased to dangerous levels, putting many species at danger of becoming extinct. In 1960 a host of African nations asked for worldwide cooperation to control illegal trade in wildlife.

It was not until 1973, however, that the Convention on International Trade in Endangered Species of Wild Fauna and Flora (CITES) was signed into law by a host of nations to protect wildlife from illegal trade. On July 1, 1975, CITES became operational, and at last count (July 2009), 175 countries had signed as members. CITES is essential for conservation and protection of endangered species, and this treaty openly explains why conservation is critical: "Wild fauna and flora in their many beautiful and varied forms are an irreplaceable part of the natural systems of the earth which must be protected for this and generations to come; [we are] conscious of the ever-growing value of wild fauna and flora from aesthetic, scientific, cultural, recreational and economic points of view" ("Convention").

While CITES has thus far been very successful, a number of powerful countries have pushed for an altered interpretation that facilitates trade in endangered wildlife in spite of the fact that the CITES preamble clearly specifies that contracting states must commit to implementation and enforcement as the document states. These nations take the view that wildlife will have a chance to survive only if wild animals "pay their way."

CITES Basics

Species listed by CITES are placed in one of three appendices:

CHART I: Species threatened with extinction.
PRINCIPLE: No commercial trade is allowed.
DOCUMENTS NEEDED FOR INTERNATIONAL TRAVEL OF CHART I SPECIES: import and export or reexport permits.

CHART II: Species not necessarily threatened with extinction but requiring controlled trade to avoid endangerment.
PRINCIPLE: Commercial trade is allowed for these species when legal and sustainable, but it must be regulated and monitored
DOCUMENTS NEEDED FOR INTERNATIONAL TRAVEL OF CHART II SPECIES: export permit or reexport permit. Some importing countries also require an import permit as an added safety.

CHART III: Species that a particular country has asked CITES to help protect.
PRINCIPLE: Commercial trade is allowed for these species, but it must be regulated.
DOCUMENTS NEEDED FOR INTERNATIONAL TRAVEL OF CHART III SPECIES: export permit from the listing country. A certificate of origin is required when individuals from these species are exported from any other country.

CITES defines trade as import, export, reexport, and introduction from the sea (in cases where a marine animal is caught at sea and brought into a country). As already mentioned, an export permit is required before a CITES species, body part, or derivative is sent out of a country. For live primates listed in Appendix I, additional documentation from a scientific expert is required stating that the recipient is suitably equipped to house and care for this imported primate.

Some people who read CITES Appendix II believe that this list of species can be traded, but this is not the case. First, to be traded, individuals must have documentation proving that they have been legally obtained. A number of countries specifically protect nonhuman primates; some even ban all wildlife exports. For example, Brazil holds the view that its wildlife is too precious to be captured and sold abroad. Consequently, primates from Brazil cannot legally be traded, even though they only appear in Appendix II of CITES. Brazil's position is both progressive and highly commendable.

Second, trade in individuals from species listed in Appendix II is permitted only if this trade does not diminish population numbers. This means that sufficient information must be collected to estimate the population size in each country where these species live, and a management plan must be in place to protect these species before approving exports for a commercial market. However, some countries allow exports for noncommercial scientific purposes or breeding programs to enhance endangered species yet fail to meet strict requirements for data and management plans. In many cases, adequate data on species numbers do not exist and/or the country of origin has no management plan, so trade is not permissible. In particular, sufficient data are not available for smaller primate species, for instance, for mouse lemurs who live in Madagascar. In these common scenarios, trade in species from Appendix II is not permitted.

Species listed under CITES are reviewed every two or three years; at that time, member states meet. During these meetings, species can be moved from one appendix to another or included or excluded from the CITES lists. As it turns out, more and more species are threatened or endangered—largely by human activities—and must be added to the CITES lists.

Safeguarding CITES species means protecting individuals not only when they are alive but also when they are dead, including body parts and derivatives. Some humans consider primate flesh to be a delicacy and/or believe that primate parts and derivatives have medicinal benefits. U.S. immigrants from western Africa are often caught smuggling smoked primate meat, for instance.

CITES uses the term *specimen* to help protect body parts of endangered species from trade. The treaty defines specimen this way:

- any animal or plant, whether alive or dead;
- any readily recognizable animal part or derivative.

Whether primates are dead or alive, their body parts and derivatives are covered by CITES and therefore protected from international transport. Permits are required to export either individuals or body parts from listed species.

CITES is designed to protect species and does not include animal welfare issues that may arise when transporting live animals. However, it does outline special guidelines for transporting protected animals. These guidelines are far more detailed than the ones written by the International Air Transport Association (IATA), for example, and there are strict rules for transporting primates internationally. Permits are always required.

Primates and CITES

From the very start, all nonhuman primate species were listed in CITES. The great apes (excluding human beings) and a few other primate species were listed in Appendix I while all other species were listed in Appendix II. One primate listed in Appendix I, the golden lion tamarin, had been traded almost to extinction because of its beautiful golden fur, which was much sought for muffs used by ladies to warm their hands in the nineteenth century. Fortunately, Appendix II in the original CITES document listed "Primates spp," meaning all species of primates not included in Appendix I. This inclusive listing remains particularly important when a new species is discovered: Appendix II already covers any new species that we may locate.

CITES lists only species that are traded. A primate may be highly endangered, but if it is not normally traded, it will only appear in Appendix II.

The IUCN (International Union for Conservation of Nature) list is entirely separate and different from CITES. IUCN lists 415 species of nonhuman primates, two of which are already extinct: the Jamaican monkey and the large sloth lemur. Thirty-seven primate species are critically endangered, which means there is extremely high risk of extinction in the wild. The black-faced lion tamarin is an example of a critically endangered primate. Eighty-six species of primates are endangered and at very high risk of extinction in the wild. Examples of endangered primates are the West African red colobus and the Zanzibar red colobus monkey. Seventy-eight species are vulnerable and at high risk of extinction in the wild, including some of the langurs and the Diana monkey. There is not enough information to judge the status of 56 primate species. This is the case for a number of the sportive lemurs from Madagascar and some of the marmosets from South America. IUCN notes that 23 species are not endangered and 133 are of least concern among primates, for example the crab-eating macaque.

CITES also safeguards endangered species by protecting habitat (flora) and may take an even greater role in this task in the future. For example, a recent interview with Erik Patel (http://www.mongabey.com), who is working to protect Madagascar primates, pointed out that logging rosewood (and other precious hardwoods) greatly endangers primates (and other species). He explained that CITES protection for these trees would help stop the destruction of forests, which would also help safeguard Madagascar's endangered primates.

The Primate Trade

Worldwide, primates can be and are legally traded for research, testing, and producing medicines. For example, in the first half of 2009, the United States imported 12,197 primates for research and testing. Luckily, there is no need to use primates for research or testing. For example, we are able to cultivate human (or other) body tissue for research, and no animals are exploited or suffer.

Many countries have banned export of primates who are taken from the wild. Others are phasing out the exploitation of wild-caught primates. Unfortunately, as a result, some countries have begun to farm primates for scientific use. Needless to say, this is no better for the individual primates who are condemned to lives of extreme suffering in laboratories.

For decades primates have been popular in the pet market, though today many countries do not allow primates to be imported for this purpose. The European Union (EU) banned this type of import and national trade more than twenty years ago. The EU permits only marmosets to be traded as pets and even then just if they have been bred in captivity. The United States does not allow imports for the pet market but unfortunately does not ban trade within the country. This means that any primate smuggled into the U.S. can be traded as a pet. There was a call in February 2010 for a new federal law to ban all U.S. trade in primates, so hopefully this will soon pass and close this damaging loophole.

Unfortunately, smuggling continues both because it is lucrative and because consumers remain ignorant of the problems they cause when they buy primate pets. Customs officers too often find primates with their heads wired to prevent them from screaming, wrapped up and hidden in carry-on luggage in an attempt to conceal their presence. These primates are confiscated, and the smuggler usually receives a fine of several thousand dollars. Sadly, it is often impossible to return these primates to their wild homes. Other options, according to the law, are euthanizing the unfortunate animals or placing them in a zoo with other primates of the same species as part of a breeding group. In such cases, the individuals typically remain the property of the country where they were confiscated and cannot be placed anywhere except a zoo.

Barbary macaques, which live in the forests of the Middle Atlas Mountains of Morocco, provide a sad example of the way such illegal trade endangers species survival. It is estimated that up to 60 percent of the young macaques born each year are taken from the wild. Due to a lack of international law enforcement, they are sold in local markets, then hidden in cars and smuggled into Europe on ferries bound for Spain. If this smuggling continues, there is considerable risk that this species will completely disappear from the Moroccan wilds within the next decade. While custom officers and police in Spain are struggling to control this smuggling, too many vehicles arrive daily, especially during summer holiday season, to apprehend smugglers and rescue these unfortunate macaques.

Health Regulations

Primates can carry diseases dangerous to humans. Therefore, strict veterinary rules apply when transporting a primate from one place to another, and prior health permits are needed. This health risk offers yet another reason not to capture primates in nature nor keep them as pets.

How Citizens Can Help

All of us help protect nonhuman primates when we do not buy them as pets, or their body parts, or any of their derivatives. Additionally, we must pressure our governments to ban all trade in primates and actively work to protect both primates and their natural habitats.

Conclusion

CITES, when implemented and enforced, has proven to be an important tool for protecting species endangered by commerce and preventing extinctions caused by trade. For CITES to remain effective, strong national legislation—including heavy fines and imprisonment—is needed. In addition, customs officers must receive thorough training in CITES enforcement, including information on smuggling methods and document fraud.

3
—

Friends of the Earth Malaysia

Phaik Kee Lim

Mr. S. M. Mohd Idris, the founder of Sahabat Alam Malaysia (SAM, or Friends of the Earth Malaysia, FOEM), is very aware that the country's flora and fauna need protection. SAM is a grassroots organization that attempts to correct problems affecting humans, nonhumans, and the environment. I have a particular interest in issues pertaining to wildlife, specifically animals affected by human encroachment, greed, ignorance, and indifference. Because of my passion for animals, both wild and tame, I am committed to working to protect them.

Nonhumans are always at the mercy of humans. Some countries eat dogs while others hunt wildlife or train animals for the circus; people everywhere brutalize defenseless creatures. In my difficult work, I am often motivated by one of the sayings of the Buddha that I read in a magazine, *Animal Citizen*: "As much as you value your own life, you must also value the lives of others."

I write letters to the media and correspond with local authorities about animal issues. I also create and send petitions, action alerts, and information on problems affecting the environment and wildlife, such as animals in entertainment; animal experimentation and vivisection; the wildlife trade; poaching; zoos; farm animals raised for milk products, eggs, and meat; the pet trade; invasions by exotic species; the aquarium fish trade; the suffering of aquarium fish; and any other forms of abuse or cruelty to animals.

I believe that animals should not be killed for human vanity, for food, for luxury or decorative items, for aphrodisiacs, or for entertainment. I detest angling for pleasure, a sport where fish suffer to entertain humans. Birds should never be caged to please us. They value their life and freedom as much as we treasure ours. I strongly discourage people from buying animals from pet shops because this contributes to the cruel trade in wild-caught animals.

Our wildlife suffers tremendously here in Malaysia. The *Malay Mail* (December 17, 1988) reported that a mother monkey—a spectacled or dusky leaf monkey—was wounded by hunters. She was discovered dead but still cradled her infant on her lap. Estate residents speculated that the wounded mother had scurried up toward safety, clutching her baby, after hunters shot her. Another monkey was found dead just two hundred meters away, also shot. It appeared that the mother monkey hid behind a rubber tree, where she used one of the cups (set out to collect rubber) to offer her baby a drink. There she remained, waiting for death. It is possible that she cradled her baby with such protective vigor—no doubt intensified by pain and fear—that she suffocated her little one. After reading this story, I wrote a letter to the editor, calling for a ban on such hunting.

SAM is working with other coalition groups to protect long-tailed macaques from being shot by people who feel these primates are overpopulated and therefore constitute a nuisance. Sometimes these macaques forage for food in refuse bins, yards, gardens, or even human homes. They pluck ripe fruits from trees, snatch grocery bags from pedestrians, and help themselves to whatever food they can find. When people complain, the wildlife department sends officers to investigate, and they often shoot the monkeys to solve the problem. This is certainly not a happy resolution for the macaques, especially since they are sometimes merely wounded. Shooting monkeys leaves many orphaned and wounded; some die quickly while others suffer much before death. For example, on October 14 of 2003, the media reported a mother macaque with two gunshot wounds in her chest, who was searching for her baby in a car park off Jalan Chow Kit in Kuala Lumpur.

SAM works to ensure that long-tailed macaque populations are controlled through sterilization—I only wish we could do the same for human overpopulation! Through sterilization, macaques will have a chance to live without being shot, and yet their numbers will be kept under control so that they do not aggravate humans. In cases where monkeys in urban areas have lost their original homes, and their chances of returning to a natural habitat are nonexistent, SAM advocates for humane methods of removing these unfortunate individuals.

I do not attend circuses, theme or wildlife parks, or zoos. Even many years later, I cannot forget an incident that I witnessed at the Royal Indian Circus when I was a child. Everyone was looking forward to watching the much-acclaimed pair of counting dogs—small ones who could read and count. The audience was asked to give the dog two numbers to add. The numbers were written on a whiteboard, and the dogs were required to pick the correct answer, from one to ten, written on cards. One of the dogs picked a wrong number, and the trainer slapped it severely across the face. Traumatized by his anger and viciousness, I was reduced to tears. I could only imagine how the dog felt.

Circus animals, including primates, are all abused in training. People who go to circuses usually never think about how these prisoners have to live, the way they are chained, caged, or tethered for most of their lives. They are trucked all over and forced to perform unnatural acts on demand to entertain audiences. If they do not do what is expected, angry circus men punish them. Our self-indulgence causes all of this suffering. Please do not attend circuses where animals are caged or forced to perform.

All of us should report to the authorities when we see animal abuse. In 2005, when I was on an investigative trip to the Kuala Lumpur Bird Park, I discovered two orangutans who were being forced to perform tricks. A young orangutan was slow to respond, and I saw the trainer pinching her under her forearm. As a member of SAM, I brought this to the attention of the office of Trade Record Analysis for Flora and Fauna in Commerce (TRAFFIC) in Petaling Jaya. Consequently, the wildlife department was forced to investigate how these primates had been obtained. They conducted DNA tests, and as a result, these smuggled orangutans were seized and returned to their home country—Indonesia. The show at the Kuala Lumpur Bird Park has since closed. When we see something cruel, we should always pause to consider ways we can stop such behavior.

A famous resort in Malacca continues to force orangutans to perform for guests. Animal acts are neither entertaining nor educational; they devalue both the exploited animals and the uninformed viewers, who fail to recognize the deprivation and humiliation that these animals suffer. Animal shows continue to exist because of human indifference and ignorance. We can change this, but we have to stop supporting (with our entrance fees and silence) forms of entertainment that exploit and harm animals, including zoos and circuses.

SAM also works internationally to protect wildlife. For example, our group partners with primate organizations such as the International Primate Protection League (IPPL). In 2002 IPPL notified SAM that four gorillas had been imported to Malaysia's Taiping Zoo. Both SAM and IPPL were doubtful that these primates had been acquired legally. It is illegal to catch primates in the wild and transport them to Malaysia. IPPL had reason to believe that these four gorillas had been exported from Nigeria with false documents claiming that they had been born in captivity (at a newly opened zoo). IPPL also suspected that the gorillas had been illegally caught in Cameroon, then smuggled across the Cameroon/Nigeria border. We were all even more suspicious when smugglers were apprehended and two baby chimpanzees were confiscated on the border between Cameroon and Nigeria.

SAM, in support of IPPL and other international wildlife organizations, sent letters of protest and action alerts to the Malaysian Environment Ministry, demanding that the Taiping Four—as they were called—be returned to their home country. Sustained international cooperation among activists was a great success for the Taiping Four; the gorillas were sent home in 2007.

In August of 2005, SAM received e-mails revealing that the Thai and Kenyan governments were planning to capture three hundred wild animals from Kenya for the Chiang Mai Night Safari Zoo in Thailand. SAM is always opposed to the capture of free, wild-roaming animals for imprisonment and exploitation, whether for science or entertainment, whether they are primates or pachyderms. How would these wondrous animals be moved from wide-open savannas, and how would they adjust to comparatively tiny, barren pens? What psychological, emotional, and physical damage would they suffer as a result of their capture and confinement? What worthy purpose would their capture serve?

In cooperation with other international wildlife groups, SAM contacted Thai Prime Minister Thaksin Shinawatra, strongly objecting to importation of Kenyan wildlife for the zoo. Organization members sent letters to the Kenyan president and minister for the environment urging them to keep and protect their wildlife as part of the magnificent national heritage of all Kenyans. SAM action alerts went out to many wildlife groups, calling for letters of protests to local Kenyan and Thai embassies. If enough people raise their voices in protest, officials will listen. The animals cannot speak for themselves. That is why, through SAM, I try to speak on their behalf.

In February of 2006, SAM received news from Edwin Wiek that Thailand's Safari World had more than 100 orangutans, who are indigenous to Indonesia. In September of 2003 and July of 2004, a joint Thai/Indonesian government inspection of Safari World discovered 115 orangutans confined in squalid, cramped conditions. Many were not registered with the authorities. Spokespersons for Safari World claimed that the park had produced these primates through a breeding program, but the disproportionate number of young individuals suggested otherwise. As experts awaited DNA tests, many suspected that Safari World's orangutans had been smuggled in from Borneo and Indonesia, rather than bred in captivity in Thailand. These smugglers, and all who kidnap wildlife from their native jungles, are decimating already-diminished populations of wild primates.

To prevent any further kidnapping, SAM joined forces with Sean Whyte from the United Kingdom's Nature Alert, as well as the larger global community of animal welfare and conservation groups, to condemn the illegal capture and transport of wild animals and demand the repatriation of these orangutans to Indonesia for rehabilitation and release back into the wild. It was my privilege to coordinate a letter-writing effort to the president of Indonesia, the minister of the environment, and various Indonesian NGOs (nongovernment organizations), urging them to demand the return of their orangutans. SAM also asked Indonesian authorities to pressure CITES officials to impose sanctions on Thailand, as well as Cambodia, Malaysia, and Saudi Arabia, for allowing orangutans to be smuggled across their borders. Letters and alerts were sent to the Thai prime minister and the director general of national parks requesting that they return illegally captured orangutans to Indonesia. These letters were followed by an action alert, asking members and other interested parties to send similar letters to both Indonesian and Thai authorities.

I continue to work to increase awareness by writing articles and letters to the media to educate citizens about what is happening with Malaysia's wildlife and any other exploited or abused animal in neighboring countries. Resources are limited at SAM; we do not have the means to establish an animal sanctuary. So we speak out. When we feel strongly—as we do about animal rights—we speak up and write letters to encourage change. I have found that the pen is mightier than the sword. We must all refuse to support industries that exploit animals and be willing to write letters of protest when animals are treated cruelly—not just primates but all animals. It will be a sad day when any of Malaysia's endangered species go the way of the dodo bird.

If people understood how much animals suffer—if they had seen what I have—they would not support animal industries. If we do not understand what our money causes, if we do not understand the inevitable suffering that results from paying entrance fees, buying animal products, and choosing what we will eat and wear, then animal exploitation will continue as other creatures are exploited for food, clothing, and entertainment. Where there is a demand, there is always a supply. Killing will only end if we stop supporting organizations that profit by taking advantage of animals and instead start speaking out against exploitation of nonhumans.

4

—

Looking Up, Counting Down

Noga and Sam Shanee

We just returned from a week of taking a mammal census in the cloud forest of La Esperanza in the tropical Andes of northeastern Peru. We found three groups of monkeys in one very lucky day. All the groups were healthy—chubby with shiny hair—and all the females carried babies. They came close to us and seemed very happy and proud to be photographed.

We were counting yellow-tailed woolly monkeys, one of the twenty-five most-endangered nonhuman primates in the world. They are endemic to a narrow band of the cloud forest on the eastern slope of the Peruvian Andes between sixteen hundred and twenty-four hundred meters above sea level. Yellow-tailed woolly monkeys are big, with males weighing up to fifteen kilograms (thirty-three pounds); their bodies are bulky and muscular, and females are only slightly smaller than males. Their dark red fur is long and thick, and when caught at just the right angle by the sun's rays, their coat has a metallic shine. They have a triangular white patch around their mouths that makes them easy to distinguish from other monkeys. They are named after the yellow hair on the lower side of the tip of their tails, but much more pronounced is the males' orange scrotal tuft, which can be fifteen centimeters (about six inches) long. The males seem very proud of their tufts and happily show their backsides to scientists and cameras. According to scientific estimates, there are less then a thousand of these animals in the world.

Unlike other primates and in spite of the fact that woolly monkeys are hunted in many areas, they have not developed a fear of humans. When they hear people, either walking quietly in the forests or loudly parading through the jungles, they come to investigate what is happening. First they show their complete disapproval of human existence in their territories by screaming and mooning the visitors with their tufts. After a minute or two, when they see that we are not impressed, they go back to their business of eating and monkeying around, but mostly they watch us. Apparently we are very interesting creatures, and they study us carefully.

On this same trip, we also saw coatis and night monkeys; we heard the calls of two pumas and saw the tracks of many wild animals living in Peru. This vibrant and colorful forest is one of the most beautiful and special places we have ever been.

Yellow-tailed woolly monkeys are not only a flagship species for conservation in the precious cloud forests of the tropical Andes, but they also illustrate the situation faced by many other endangered species around the world. The forests of La Esperanza are important on a global level because they form part of the catchment basin for the mighty Amazon River. Constant, heavy rainfall fills local rivers, which flow rapidly down the mountains to irrigate the Amazon forest, the largest rain forest on the planet and therefore Earth's most important reservoir of fresh water and air.

Throughout the week, while we were taking the census, we heard the buzzing of chain saws. As we stood under a group of woolly monkeys, admiring their shining fur, all we could hear was the ear-piercing noise of chain saws, punctuated by the sound of falling trees. The monkeys have become so used to the roar of the chain saws that they are oblivious. The saws start at eight in the morning and continue almost until sunset. The inhabitants of La Esperanza are cutting the forest. Some of the most valuable timber is sold on the illegal market for a fraction of its true worth.

But this logging only destroys a handful of timber species and leaves most fruit trees, and many other trees, standing. The real problems are the legal logging operations that build roads, opening up new areas to settlement, and cut many more trees then the individual illegal loggers, and the widely promoted cattle ranching. Local ranchers, supported by regional and national governments, clear-cut ever-increasing areas to pasture herds of cattle. Thousands and thousands of hectares of forests are disappearing each year, and these forests are the habitat—the only home—for this rare and beautiful species of woolly monkey.

The inhabitants of La Esperanza are not native to this land—they do not count as Indigenous, which we Westerners too often picture as the people most connected to the land—the forests' fiercest protectors. They are mestizos—people with mixed Peruvian and European ancestry—and they are migrants. They come from other areas of Peru, where human activities have made the land totally infertile and the water scarce. Having destroyed their previous home, they migrated and built a new settlement, which they call La Esperanza, meaning "the hope" in Spanish. Now they are clear-cutting the home of the woolly monkeys, and these residents cannot simply migrate to another area of South America. If the clearing continues for a few more years, La Esperanza will not have enough forest to support a viable population of yellow-tailed woolly monkeys.

La Esperanza is merely a carbon copy of activities all around South America. Not only the woolly monkey but thousands of other species are being displaced by our consumption of wood products, by our cattle industry. It is probable that yellow-tailed woolly monkeys will be extinct, at least in the wild—in their homeland—in just a handful of years. Moreover, deforestation in such mountainous regions brings dramatic climate change and destroys the quality of the soil. It is likely that La Esperanza will soon be unsuitable for human habitation; after destroying wild animals who depended on these forests, humans will merely move on to consume another forest and another community of wild animals.

The number-one cause of global climate change, the number-one cause of tropical deforestation, and one of the greatest forces in the depletion of potable water is our diet—raising cattle for milk and beef (FAO 2006). In La Esperanza, cattle ranching is the greatest threat to the cloud forest and definitely poses the most critical crisis for the endangered woolly monkeys. Why are the inhabitants of La Esperanza cutting down these forests? Don't they know that it is bad for them and everything that lives? Don't they care about the monkeys? Are they too poor to have other options? Are they uneducated—unaware of what they are doing to themselves, the animals, and all of us? Do they like meat so much that they are willing to sacrifice all the forests and animals who live there for the sake of a beef burrito?

The people of La Esperanza are very aware that cutting down the cloud forest jeopardizes their future well-being. They are worried about the woolly monkeys, all the other monkeys who live in these forests, and the forest itself; they would not deliberately harm any creatures of the forest. But they are poor. They are so poor that most of them cannot afford to eat the beef that comes from the cattle that graze on their land once the trees are gone. They cannot afford to drink the milk that these cows produce. It is we consumers in the West who buy the milk and the meat. It is we who cause the cutting of the forests of Peru, and it is we who are driving the yellow-tailed woolly monkeys to the brink of extinction when we go shopping. We can—and we should—make different choices: for the sake of the forests, or the woolly monkeys, or the cattle, or the people of La Esperanza—or for all of them.

We traveled to La Esperanza two and a half years ago as part of a short research project on conservation and the yellow-tailed woolly monkey. After witnessing the devastation and realizing that these lively woolly monkeys were just a tail length away from extinction, we raised the necessary funds, left our separate homes in England and Israel, and settled together in Peru. We soon founded Neotropical Primate Conservation, an NGO (nongovernment organization) dedicated to the preservation of South and Central American primates generally and the yellow-tailed woolly monkey in La Esperanza specifically.

Our arrival in La Esperanza was a big event for the locals; most of them had never had any relationship with "gringos." They knew enough to be very suspicious, however. They believed we came to steal their children, steal their lands, and even worse, steal and sell their monkeys as other Westerners and Western-trained scientists have done in South American forests. They had never been invited to join a conservation project developed specifically for their forests—their local flora and fauna. They were not even sure exactly what conservation meant. In spite of a history of white exploitation in South America and their initial fears, the villagers were very welcoming, and they donated a house and an office for our use.

We soon discovered that—in spite of this area's unique biodiversity, in spite of the fact that these jungles are home to some of the most endangered animals in the world, and in spite of the fact that this area is understood as critical to the survival of the planet—governments and international NGOs were (and are) failing to preserve these special forests. The Peruvian government, motivated by corporate interests such as those of mining, oil, and the meat and dairy industry, encourages the destruction and unsustainable exploitation of the country's forests. The United States, European governments, and megacorporations promote—through various Peruvian government ministries—ever-growing numbers of cattle and mining operations, insisting that these are the path to development and riches—the surest and quickest path out of poverty.

But the only ones getting rich are the corporations. The local Peruvians who carry heavy chain saws in the hot sun are paid so little that they must work constantly just to survive, and of course they must keep cutting one forest after another. International companies that pass through the area have left behind nothing but a legacy of unfulfilled promises and environmental devastation.

International NGOs are also far behind schedule in protecting these forests and their many inhabitants—including the local people. They are rarely in contact with local communities. Immigrants and hunters have swarmed into these few established conservation areas, and there is little support or funding for authorities who might otherwise be willing to protect these precious preserves. Yet foreign money has poured into printing shiny books for a population that is largely illiterate and holding fancy meetings to which local populations are not invited. Many well-designed maps and reports describe—in great detail—projects that make very little difference in Peru—for the forests, for the nonhumans who live there, or for native people.

We, as concerned donors to conservation projects, must find a way to make sure that our money is spent where it is most needed and not get trapped in the long corridors of gigantic NGOs. One of the best ways to ensure that your money and effort go where need is greatest is to donate to smaller organizations, which don't have large donors and therefore rely on small contributions to survive. Small organizations with little money to waste on overhead often do the groundwork. You can also visit these NGOs or volunteer at projects that seem particularly interesting to you.

Before we arrived, many of the people in La Esperanza did not know that the busy primates in their forests were, in fact, endangered yellow-tailed woolly monkeys. They hunted the monkeys if they thought they could sell them for a few soles (nuevos soles are Peruvian money; approximately three soles equal one U.S. dollar) and never thought twice about cutting down the forest. Though this monkey is one of the most endangered primates in the world, living in one of the most biodiverse areas on the planet, the locals must fight just to get help, or at least a little direction, to protect their precious local ecosystems.

With new information, the situation changed quickly. Even though some locals remain suspicious and hostile (for good reason, given the track record of outsiders), growing groups of people keep in close contact with us and are changing their relationship with nature, the forests, and the monkeys. Without any outside pressure, these broad-minded communities—when given new information—voluntarily stopped all hunting of endangered wildlife, and now they enforce this ban with their own communal organizations. We receive daily visits from both locals and people in neighboring communities who want to conserve their lands. Some locals are interested in creating reserves on private lands that are ten to twenty hectares (about twenty-five to fifty acres); others want to create reserves on communally held lands of two to three thousand hectares (five thousand to seventy-five hundred acres).

Many individuals and institutions, such as churches and schools and communal organizations, approach us for more information about their forests and wildlife. For example, the Ronda is a national network of locally organized vigilance groups that live in remote rural areas, where police and regional or national authorities have little presence, and these groups are working with us to protect forests and endangered species.

Among the many people who approach us from this host of different institutions, people from churches are some of the most interesting and promising allies for conservation. There are many churches in the area, representing many different denominations, and the vast majority of the people are devoted churchgoers. For example, the village of La Perla del Imasa, which has a population of about fifty families, has at least six different churches. Although Latin American churches have quite a bad historical reputation in relation to human rights and the environment, what we have experienced in La Esperanza is quite different. Even church leaders can be inspired by stewardship toward nature as required of humanity by the Book of Genesis. They see environmental deterioration caused by humans as the destruction of God's creation—creation that God called "very good." Churches, like other institutions in the area, have never received any practical or spiritual guidance about the way to practice more responsible environmental stewardship. We see the churches as potentially very influential environmental educators for both children and adults; these institutions and their leaders are deeply respected within communities and can reach the largest number of people in some of the most remote areas.

Like ourselves, the people of La Esperanza are immigrants and have much to learn about the forest and local wildlife. They also want to know more about the larger world, and they are very proud to learn that their forests have such remarkable biodiversity—and are critical to the planet's survival. They can see the destruction that they have inadvertently caused, and they want to know what they can do to save this gift of green.

We have seen yellow-tailed woolly monkeys on many different occasions: We found them happy and healthy in primary forests and trapped in small patches scrambling to find enough food to survive; we saw them boldly displaying their scrotal tufts to passing humans; and we saw them die at the hands of humans who captured and tortured these endangered individuals. Throughout our work, we constantly swing between total pessimism and dark prophecies for the species' extinction to optimism about their survival. Our hope rests in dedicated local communities and individuals that we know have the strength and determination to reverse the seemingly doomed future of these monkeys. La Esperanza, as its name suggests, is our hope.

5

International Primate Protection League

A Wonderful Life

Shirley McGreal

My husband, John McGreal, had been transferred from India to Thailand, so I was in the Bangkok International Airport, collecting my boxes and cases, when I spotted a shipment of caged infant monkeys. The babies were snow white and had been packed into long crates divided into several compartments with wire windows on the front. Each compartment contained two to three babies looking at me as if they were seeking help.

Those little monkeys were destined for a foreign country, and I wondered why they were leaving their homeland. Later, I learned that they were stump-tailed macaques and would have reddish hair when grown up, though they are born white. I also learned about the atrocious suffering of laboratory monkeys, and I feared for those little traveling macaques.

I had no background with nonhuman primates. My doctorate is in educational history, and I wrote my dissertation on the education of Indian Rajput princes. I expected to return to the United States and teach at a university.

But those baby monkeys led my life in a totally different direction. I became absorbed with primates. They were everywhere in Bangkok. They were on sale at local markets, such as the huge, sprawling Chatujak. Locals kept baby gibbons, monkeys, and even slow lorises as pets. "Photo touts" on the beaches used gibbons; tourists paid to have their photos taken with a Polaroid camera while holding a clingy baby gibbon. What the tourists did not know was that the mother gibbons had been shot in the jungles of Thailand to bring their babies into captivity for human amusement. That's why the babies clung to anyone—baby gibbons normally spend twenty-four hours a day attached to the furry bellies of their mothers.

Of course I wanted to help, so I looked for an organization assisting primates. I couldn't find one, so I decided to form one myself. International Primate Protection League (IPPL) started in 1973 with one member, me! But gradually word spread, and we attracted a few Thai members and a few supporters from other places around the world. One Western lady living in Thailand enrolled twenty-five of her friends! Now IPPL has thousands of supporters all over the world.

I read everything I could lay my hands on to educate myself about primates. After I read *The Apes* by Vernon Reynolds of Oxford University, I wrote him a letter explaining my plans. To my delight and surprise, he answered me. He had never heard of me, of course, but he did not ask, "Who are you, and what are your qualifications?" This terrific man wrote back saying that the group I planned to form was a great and long-overdue idea, and that he would like to help me and join our committee. To this day, Dr. Reynolds remains a dear friend and advisor to IPPL.

We issued our first newsletter in 1974. It was typed, photocopied, very unprofessional, and nothing like the full-color magazine we publish today.

Soon I was joined in my work by Ardith Eudey, a graduate student studying wild monkeys in a Thai wildlife sanctuary. (Ardith went on to earn her doctorate in anthropology and still works closely with IPPL.) Her research accommodations were rough, so she took time off to stay with me in Bangkok. Together we uncovered a ring of smugglers shipping baby gibbons from Thailand to Canada, then on to a California laboratory for medical experiments. One shipment of ten babies arrived with six dead—one with a shotgun pellet lodged in his brain. No doubt he caught this bullet accidentally when his mother was shot.

We went after the smugglers. Thai wildlife authorities gave me a copy of one legal export permit for "eighty mynah birds." Then I had a lucky break. I went in pursuit of the same documents from the Canadian dealer, Kenneth Clare of the Ark Animal Exchange. I arrived, with no notice, on his doorstep, and he welcomed me! We had a long chat, and he willingly copied all the documents related to his primate shipments. We soon discovered that the permit for eighty mynah birds had been altered—ten heads of white-handed gibbons had been added to the list of exported individuals. We could tell that the words had been typed on a different typewriter. We had discovered how the smugglers operated!

We were very excited and soon reported our findings to Thai and U.S. authorities. Two crooked Thai customs officials were fired, but the United States never prosecuted the case, even though the shipment clearly violated the Lacey Act.

In 1974, on leave in the United States, I obtained from the U.S. Fish and Wildlife Service documents showing that large numbers of gibbons had been imported from Singapore—a nation that has no wild gibbons. I headed for Singapore, where I pretended to have twenty gibbons that I wanted to transport from Thailand to the United States. "No problem!" I was told. One man explained that he would put them in false gas tanks under trucks traveling from southern Thailand via Malaysia to Singapore. Another said he would put them on a coastal freighter and bring them by sea from Thailand to Singapore.

Animal traders do not like the light of day. They work in anonymity and shadows, where some become multimillionaires. I returned to Thailand and wrote an article on what I called the Singapore Connection. A Singapore paper published my essay, and the Reuters wire service carried the story around the world. Because of that article, no more gibbons reached the United States from Singapore. More than thirty years later, IPPL continues aggressive campaigns to end the international primate trade.

In 1977 I settled in Summerville, South Carolina, with my husband. Summerville was then a lovely, quiet little community nicknamed "Flowertown in the Pines." The town is lovely year-round but especially in the spring, when azaleas, wisteria, and dogwood bloom. At other seasons, we enjoy camellias and sweet-smelling magnolias. Flowertown provides a perfect climate for IPPL's primate sanctuary: weather that is generally hot and humid. Sadly, the nice climate drew developers, and the town is not what it used to be, but the gibbons are happy here, and so we are, too.

Humans harm nonhuman primates in many ways. We cut down their forest homes. In many countries, people eat other primates. This was less of a problem when Indigenous people killed animals just to feed their families, but the construction of roads deep into forests—usually for logging—and the easy availability of guns have helped poachers kill apes and monkeys—and many other species—in large numbers for large populations in city markets. I visited one such market in Yaounde, Cameroon—a ghastly spectacle. One man drove up in a Mercedes to buy a monkey for dinner. Some of this bushmeat ends up in European cities such as Madrid and Brussels, where large numbers of African residents now live.

Monkeys are also exported for the pet trade. While the United States bans such imports, we allow breeders to sell monkeys. Nobody would buy such pets if they saw the harrowing scene of a baby monkey being kidnapped from his or her loving mother. In any case, baby monkeys do not make good or happy pets. Their owners, who pay thousands of dollars for a monkey, often make them wear diapers and treat them as if they were human babies. Primate vet care is difficult to find, and many of these babies die for lack of medical attention. Such monkeys are also likely to become aggressive, and many owners extract their sharp canine teeth. Mature monkeys—no longer cute or easy to manage—are often sent to roadside zoos, where they live out their lives in small depressing cages, most often alone. Very few primates reach sanctuaries because most of them are full with long waiting lists.

Most frequently, exported monkeys become laboratory animals. Thousands of monkeys are also bred in research facilities in the United States and many other countries specifically for this purpose. Primate experimentation is extremely cruel. Over the years, IPPL has campaigned to ban all primate exports around the world. One of our key strategies is to inform the citizens of export countries what happens to their monkeys in other countries; we expose the painful and often-gruesome experiments conducted at the expense of their health and welfare. The U.S. military is guilty of some of the most horrific experiments, such as radiation tests and those for biological and chemical warfare. IPPL contacts ambassadors of countries that export primates and sends out press releases to every newspaper in these countries—not just the English-language ones but also those in local languages such as Chinese or Bengali—to inform people about what happens to their nation's primates after export. In almost every issue of IPPL News, we also ask members to send protest letters and postcards to government officials.

As a result of these campaigns, India (1977), Bangladesh (1979), and Malaysia (1984) have banned primate exports. As new countries enter the monkey trade, we campaign for export bans and request that governments investigate the local primate trade. In the process, illegalities often surface, such as exporting wild-caught monkeys with fraudulent documents claiming that they were born in captivity. Because primates are expensive and in high demand, IPPL continues to fight both their legal and illegal trade. Luckily we have created a worldwide network, so we receive many smuggling tips to help in our work.

In retrospect 1981 was an exciting year for IPPL because a California laboratory using dozens of gibbons for fatal cancer experiments lost federal funding. There were more than fifty surviving gibbons, and we had to find new homes for each individual. My old friend Ardith Eudey, who was then living in California and continuing her primate work, found out that one sickly baby gibbon could not be placed and was slated to be killed in a final experiment. Ardith called me, and I contacted the laboratory director and offered funds for his care. The director said he'd rather spend the money on a one-way ticket to Summerville. He said the baby was metabolically abnormal and mentally retarded.

Mrs. Katherine Buri, a Thai member of IPPL, asked the monks at Bangkok's Wat Arun, the Temple of the Dawn, to select a name for the baby gibbon to put him under the protection of the Lord Buddha. The monks chose Arun Rangsi, which means "the rising sun of dawn."

Of course IPPL doesn't require gibbon IQ tests! So, on August 9, 1981, Arun Rangsi reached IPPL on his second birthday. He still lives at IPPL with his gibbon family. Over the years, we took in more needy gibbons, and as of April 2009, IPPL houses thirty-two of these amazing little apes. Each one brings a unique story, and it is usually a very sad one. Some were pets. Others came from laboratories or zoos. But because they have found a home at IPPL, we will make sure that their lives are happy from here on.

After he arrived, we studied Arun Rangsi's medical records. He had suffered twice from dysentery, twice from pneumonia, and had twice lost 10 percent of his body weight for unknown reasons. He had been raised alone with only a swinging wire surrogate as his mother. He weighed merely half of what a normal baby gibbon would weigh. We were amazed that he had survived. The most striking thing about Arun Rangsi was his huge, shining brown eyes.

Our new friend banged his head most of his waking hours and, as a consequence, had a callus above his left ear. A human psychiatrist came to visit and said the little ape was acting like an autistic child. He suggested I bang my head along with Arun Rangsi to enter his dark world, share his experience, and then lead him into a healthier way of living. And so I banged my head. And possibly because I entered his world, Arun Rangsi became a great little friend and companion. We worked to find him a gibbon companion and thought he would have one when a New York laboratory offered us a female named Ellen. But when John went to New York to collect her, he found that Ellen had recently been separated from her companion, Peppy. We were able to arrange for both gibbons to come to IPPL, and the next year the lab sent us a lovely three-year-old female named Shanti. Arun Rangsi still lives with Shanti. She is as laid back as Arun Rangsi is hyperactive. They make a wonderful pair and have had several gibbon children.

Beanie was also an inspirational gibbon. He came to us from the Lubee Foundation in Florida, in 1991. The foundation was founded by Louis Bacardi of the famous rum family and had amassed nearly two hundred nonhuman primates (and many other animals) for studies. Beanie had been bitten by a mosquito during the hot summer of 1990 and had developed encephalitis, which left him blind and epileptic. He was a delightful gibbon. He lived in a specially designed enclosure during the day. IPPL staff took him out to play on the grass, joined by our blind rescue dog, Bullet. Even though Beanie developed inch-long canine teeth, he never used them on his human friends. At night he sat on the couch with a favorite human and enjoyed bananas and being groomed. Sadly, the seizures continued, and a violent one took him away from us in 2004. I will never forget his courage in the face of adversity, and Beanie's gentle face remains my computer screen saver.

Every two years IPPL holds a members' meeting. People from all over the world attend. In 2002 an Asian speaker asked for a private meeting with me. He brought out his laptop computer and showed me photos of dealers holding a baby gorilla and a baby chimpanzee and also gave me a Malaysian animal dealer's business card. He told me that four baby gorillas had recently reached Taiping Zoo in Malaysia and were being kept off display. IPPL sent Dianne Taylor-Snow to Taiping to investigate, and she confirmed the presence of the four baby gorillas. These individuals became known as the Taiping Four. Our investigation revealed that they had been smuggled via South Africa to Malaysia from Nigeria and had probably originated in Cameroon. For several years, IPPL worked to have the gorillas returned to a rescue center in Cameroon. Finally, Malaysian authorities confiscated them but sent them to Pretoria Zoo in South Africa. The campaign continued, and, in November 2007, the Taiping Four were finally returned to the Limbe Wildlife Centre in Cameroon.

When IPPL learns of primate smuggling, we always conduct an investigation. Once we have identified the smugglers, we work with the nation of origin to investigate and prosecute. In one such case—the Bangkok Six—six baby orangutans were confiscated at the Bangkok International Airport. They had been shipped to Bangkok from Singapore in coffinlike wooden crates labeled Birds. There was no way to see what was actually inside the crates. The animals were due to be transferred to a plane leaving for Belgrade, but the flight was cancelled. Luckily the drugged babies woke up and started to whimper. Thai authorities thought they were hearing the cries of human babies and passed the crates through an X-ray machine. They saw the profiles of six baby orangutans and two gibbons. The animals were seized and placed with Thailand's Wildlife Rescue Centre.

IPPL again sent Dianne Taylor-Snow, an experienced orangutan caregiver, to Bangkok. She helped with the daily care of the babies while I started the necessary work of identifying the guilty parties. It turned out that the network was very cosmopolitan, involving smugglers from Indonesia, Singapore, Thailand, Yugoslavia, Germany, and the Netherlands. The ringleader turned out to be Matthew Block, who conducted a monkey-dealing business out of Miami. It is often hard to motivate U.S. authorities to investigate crimes involving nonhumans, and the case dragged on for years. We persisted, insisting on justice, and ultimately Block was indicted and went to prison for his role in smuggling these little orangutans. As for the little ones, unfortunately, four of the six babies died.

Africa has the most primate rescue centers, perhaps because so many primates are African natives. Although Chile has no native primates, some end up there, transported from neighboring countries like Peru to be sold as pets. Keeping primates as pets is illegal in Chile, and authorities confiscate them if they are discovered. Confiscated primate pets are sent to Siglo XXI, a sanctuary for South American primates in Chile run by Elba Muñoz Marin and her family (http://www.awionline.org/pubs/Quarterly/winter03/0103p13.htm).

It is known that women excel in primate studies. Dian Fossey, Jane Goodall, and others have shown that women can endure rugged conditions and make patient observations. Over the years, numerous primate rescue centers have been established around the world, and women have founded and run many of them. In Nigeria, Zena Tooze runs Cercopan (http://www.cercopan.org), a sanctuary for guenon monkeys rescued from trade. In nearby Cameroon, Sheri Speede founded Sanaga-Yong Chimpanzee Rescue Center (http://www.ida-africa.org/).

In the Republic of the Congo, Aliette Jamart founded HELP, a chimpanzee rescue center (http://www.help-primates.org/). Next door, in the Democratic Republic of Congo, Claudine Andre founded Lola ya Bonobo, home to more than fifty rescued bonobos (http://www.friendsofbonobos.org/index.htm).

As I write this essay, an eighteen-year-old volunteer is staying with us, helping to care for the gibbons. I hope that people of all ages will join the battle to save our fellow primates—and all animals. There are many things you can do, ranging from the exciting work of catching wildlife criminals to volunteering at your local shelter or adopting a pet that would otherwise be put to sleep. Even if you are housebound, you can participate in letter-writing campaigns. You can also support your favorite groups with gifts. Please be sure to check the organizations first at http://www.guidestar.org to be sure that they use their money wisely.

Some nonhuman primate species, especially great apes (chimpanzees, bonobos, gorillas, and orangutans), are particularly popular with the public, so more groups are devoted to these favored species. Gibbons are also apes, but they are not trendy like the larger ones! Tens of thousands of what I call "the plain primates" suffer terribly in laboratories: the macaque monkeys of Asia, the baboons and guenons of Africa, and several South American species such as squirrel, owl, and capuchin monkeys, as well as tiny marmosets. These unfortunate animals are used in experimental surgery; testing drugs, vaccines, and biowarfare agents; and researching infectious diseases.

My work with primates has given me a wonderful life, and I have met many great people like Indira Gandhi, Dian Fossey, and Jane Goodall, along with great gibbons like Arun Rangsi, Beanie, and so many more nonhuman individuals at rescue centers around the world. It is amazing to remember that all this started when I saw those little snow white monkeys, awaiting their fate in shipping crates in the Bangkok airport. I often wonder what happened to them, and if any of them are still alive.

Having so many rescued primates at IPPL headquarters vaccinates me against burnout. With these furry faces looking expectantly at us, we cannot become discouraged just because so many battles fought on their behalf are lost. We can't give up. I will spend the rest of my life fighting to protect our primate cousins and friends from trade and experimentation. The gibbons in my life have played a vital role in keeping me sane, motivated, and dedicated to protecting all the world's primates.

Part II

Research

6

—

Paper Lives

Michael A. Budkie

Animal experimentation is so big; we really have no adequate idea of how many animals are victimized in labs every year. About eleven hundred U.S. labs engage in animal experimentation. Tens of thousands of individuals die every day—one at a time—often very painfully in these labs. In 2007 more than 69,990 nonhuman primates were used for experimentation at dozens of universities, contract laboratories, and government facilities in the United States. It is impossible to make sense of what is tolerated legally in these laboratories: duplication, psychosis, excruciating suffering. When you spend time exploring animal experimentation—protocols, inspection and animal-use reports, grant applications, journal articles—you become an expert in pain and suffering.

What keeps this cruel system in place and motivates people to cause such long-term misery to primates and other creatures? Consider the case of Marilyn Carroll, who performs drug-addiction experiments on primates at the University of Minnesota, a project that has been funded through four separate grants, totaling roughly $1.1 million annually, and has been ongoing for twenty-eight years. Three of these grants have undisclosed amounts allocated to Carroll's salary. The fourth grant is nothing but salary, and it is funded for $138,988 per year. It would not be surprising if Marilyn Carroll is receiving an annual salary of more than $200,000. Animal experimentation is very profitable, both for the institution and those in white coats. Carroll's situation is not unique. She is just one of many researchers at the University of Minnesota, and the University of Minnesota is but one U.S. research facility engaging in animal experimentation.

It is difficult to comprehend what the lives of these victims of laboratory research are like, but I invite you to try. Medical journals list their cages as thirty-three inches long by thirty inches wide by thirty-six inches deep. To put yourself in their place, imagine that your life consists of confinement in a small enclosure, which only allows you to take one or two short steps in any direction and has just enough height for you to stand upright. You will never see the sun or breathe fresh air. Your stainless-steel enclosure is barren, designed to facilitate cleaning, containing a seat of some variety and a single rubber toy. There is nothing else to pass the time, nothing to occupy your mind. The partially open front allows you to see that others are in similar rooms nearby. You can talk to, see, and possibly smell them, but you cannot interact. The loneliness is devastating. You have no relationships with friends or family. You never even have the opportunity to touch another individual. You often feel as if you are losing your mind. Many of the others that you can see and/or hear behave as though they have lost their minds.

If you were one of these primates in a stainless-steel cage, this would be your entire life, and it would only end when you became ill, likely from septicemia or some other condition caused by experimentation. Death is your only possibility of escape from your boredom and confinement.

This is the brutal reality for monkeys in laboratories. The University of Michigan has a typical animal lab. Experiments at this facility have been under way for decades—literally consuming the lives of hundreds of primates. Two University of Michigan researchers, James Woods and Gail Winger, engage in one of the most common types of animal experimentation—drug-addiction testing. For this purpose, they subject macaque, squirrel, and rhesus monkeys, as well as baboons and other primates to decades of isolation, confinement, and agony.

According to the website of the National Institutes of Health (NIH), in just the last five years, these two researchers have squandered more than thirteen million dollars in federal grants: Winger has been federally funded since 1976, and Woods has received grants since 1971.

Almost every health record for a primate exploited in addiction experiments at the University of Michigan describes lab monkeys ripping out their hair and worse. Reports mention multiple incidents of severe self-mutilation, weight loss, and amputations resulting from lacerations. For example, University of Michigan charts show that Scallywag, one of their primates, lost weight from constant activity associated with psychologically abnormal behavior. Scallywag exhibits abnormal behavior when people are in the room. Meanwhile, Clash is listed with an unexplained 12 percent weight loss. An unnamed rhesus monkey suffered a 15 percent weight loss in just three months. This primate is also described with constant muscle contractions and as hypothermic. It sounds as though she is experiencing drug withdrawal. Yet another primate, named Data, lost 10.5 percent of her body weight in an equally short period. In 2006 Harpo is listed with four incidents of self-mutilation in just five days; he has a long history of self-destructive behavior. Eminem wears a "long-sleeved jacket due to his history of self-mutilation."

The list goes on and on. It is small wonder that these isolated rhesus monkeys have gone out of their minds. They wear a nylon jacket to cover a surgically implanted intravenous catheter, which administers addictive drugs. The catheter exits through a site on the primate's back and is connected to a metal spring arm, which is affixed to the rear of the cage, further limiting the captive's movement. It is not surprising that these monkeys can be trained to self-administer drugs since addiction is the only means left to fight their mind-numbing boredom.

Possibly the most common result of experimentation and confinement in the laboratory, at least for primates, is insanity. The laboratory environment is so artificial, so utterly contradictory to virtually every aspect of what is normal for a primate, that insanity is commonplace. Insanity is a primate's normal reaction to an extremely abnormal situation. Medical records for primates at the University of Minnesota reveal the following signs of stress and insanity:

- August 9, 2005, primate 05GP20: "Temp was up due to primate jumping back and forth wildly."

- July 26 and 28, 2005, primate #312A: "still overdosing on current drug dosage, ataxic, hypersalivating, disoriented."

- April 2, 2007, primate #312A again, generalized alopecia (hair loss): "on current drug dosage ataxic, hypersalivating, disoriented."

- August 23, 2005, primate #312E, evidence of self-mutilation: "did bite knee after observation."

- March 21, 2006, primate #45C: "extremely thin, body condition is poor, severe alopecia…bruising on top of left ankle."

- November 15, 2005, monkey #45D: "ripping hair from the armpit area and chewing on the fur, each time he would grab a tuft of fur he would vocalize."

- September 7, 2005, primate #78B: "extreme alopecia."

- September 11, 2006, #78B still has severe alopecia.

- March 22, 2005, primate #25A lost part of his tongue.

- September 14, 2004, primate #25b was overdosed.

These charts detail extreme suffering, days of agony. They describe primates that have little to look forward to other than the addictive drugs that temporarily remove them from reality. And these charts represent only what was noticed and noted.

Laboratory captivity creates psychologically abnormal individuals. Consequently, the applicability of test results is highly questionable. Clearly primates in the labs of Winger and Woods at the University of Michigan are anything but psychologically normal, rendering their experiments essentially meaningless. Furthermore, many of these unfortunate animals come from other laboratories that have also performed psychological experiments on them, such as the Medical College of Virginia at Virginia Commonwealth University (VCU), Yerkes National Primate Research Center (connected to Emory University), and the NIH itself. The Yerkes Center and the Medical College of Virginia perform drug-addiction experiments, and it is possible that these University of Michigan primates were exploited previously at one of these other facilities; they may have been severely psychologically stressed before they even arrived at the University of Michigan. Researchers at NIH also perform maternal-deprivation and alcohol experiments on primates. It is likely that any primates transferred from NIH had lost their sanity long before they arrived at the University of Michigan but were nonetheless submitted to more experimentation—presumably to learn more about humans. Primates have long lives, and many at the University of Michigan have endured psychological and addiction experiments since at least 1990. Many have undergone decades of drug addiction and psychological agony. Does any of this make sense?

For more than twenty years, I have been an animal activist working almost exclusively against animal experimentation. I've read tens of thousands of pages of inspection reports, research protocols, and health-care records for dogs, cats, goats, and primates. After a while I started to look at things not in terms of pain and suffering, or sanity and insanity, but for the quality of the information. Dealing with all of this suffering, greed, and death robbed me of emotion. I began to keep a distance, to speak in terms of press coverage and reaching people with the truth. I separated myself from the pain of other animals to deal with my pain. This is a normal response. People who witness so much suffering tend to function almost like staff in animal laboratories. If you think about the individuals who are suffering excruciating pain and ongoing despair, it is far too painful, so you maintain a distance.

If through some miracle, you don't either lose your humanity or burn out when you are witnessing such unjust exploitation, some small moment, incident, or artifact is likely to grab hold of you and refuse to let go. It may be a picture, it may be a specific animal, or it may just be something deep within that won't allow your humanity to shut down. For example, consider this sentence, which creates a searing picture in my mind that is painfully, yet inescapably, clear. "NHP was observed by LACT ripping hair from the armpit area and chewing on the fur, each time he would grab a tuft of hair he would vocalize." The psychic agony of this monkey is so abject, so pure, so complete that I feel it may drown my soul. I have been swimming in this image for days.

I did not choose to read these charts. Witnessing suffering was never my life's goal. The only reason I continue to do this painful work is to eradicate animal experimentation. When I read about puppies that have drowned in floor drains (Michigan State University), primates that are dissected while still alive (Southwest Foundation for Biomedical Research, now the Texas Biomedical Research Institute), or a female sheep that dies with two rotting lambs inside her womb (North Dakota State University), I recognize both tragedy and truth.

I can never forget that each one of the thousands of pieces of paper that I have read—documents detailing the horror of animal experimentation—actually describes the life of an individual. The lives of these victims are sketched one page at a time: inspection reports, daily-care logs, surgical records, necropsy (postmortem) reports. They record infant deaths and lives that are decades long, intelligent and sentient lives of animals like rhesus monkeys, whose laboratory existence is utterly unreal to them, yet painfully inescapable. These pages are not just statistics to be added and subtracted; these written words represent individuals whose lives matter. The only reason for any normal human to become immersed in this world of cruelty is to end it.

7

16162

Matt Rossell

She was only known as 16162, the number neatly tattooed in black ink on the inside of her thigh. Consistent with the other twenty-five hundred nonhuman primates at the Oregon Regional Primate Research Center, she was treated like a number.

Looking back at the two years that I spent at Oregon Health and Science University's (OHSU) primate center, I regret never giving 16162 a name, if for no other reason than to have someone in her stark, stainless-steel world make a trifling gesture to recognize her as an individual. She certainly deserved it and so much more. But the research community frowns on naming experimental subjects.

"They discourage special treatment," Amy told me while showing me the ropes. "It could cause problems in the data." Apparently compassion wreaks havoc in a research lab. Ironically, Amy cautioned me about naming the monkeys while feeding greasy Purina Monkey Chow biscuits to one of her personal favorites, whom she had named. Marge, the overweight crab-eating macaque, presented herself to Amy for a head scratch, an offering of trust rarely seen in research labs and universally undeserved. Amy rewarded her with a single peanut.

That seven-second interaction was no doubt the highlight of Marge's day, and it was also more personal attention than many lab monkeys receive in a lifetime. One moment of touch stood in stark contrast to hours and hours of isolated, cage-crazy boredom.

It isn't that my coworkers were mean to the animals; in my experience, outright physical abuse was uncommon. But the employees of the Division of Animal Resources got kudos for being efficient, not for being kind. Assembly-line animal husbandry is standard practice in research labs.

Take Amy, for example, whom I came to know as a mass of contradictions. On the one hand, she often raced through her cage-cleaning obligations to have time to shower treats on one room of monkeys. This simple act of kindness was outside of management expectations. Yet in her off-duty hours, she sometimes entertained herself by shooting free-ranging dogs on her property. Around the break-room table one day, she bragged of trucking lifeless dogs, which she had just shot, back to her rural neighbors. Recounting the interaction, Amy delivered her sadistic punch line with a smile: "I asked 'em, 'Do you know where your dogs are?' They said, 'In our yard.' An' I said, 'So these two dead Rottweilers in the back of my truck must not be yours then?'"

For Amy, shooting dogs who were running loose, who might chase livestock, was justifiable. I wonder if she knew those dogs by name. Another time Amy told me that she drove into a field and spun cookies to run over a pheasant with her pickup truck. She showed me the feathers.

The prospect of giving every monkey a name was far too overwhelming, but those few technicians like me who cared enough to try managed to make a few names stick. However, more often than not, names were fleeting. A cute, clever, or sometimes crass moniker might be scrawled on a primate cage, only to be washed away and forgotten the next time the cage was cycled through the autoclave for sterilization, or the name was left behind after the monkey was moved to a new location for experimentation.

I might never have noticed 16162. My boss, Carla, the head of the psychological well-being department, pointed her out when she was training me. I was her only underling, her fledgling protégé. The two of us were responsible for the mental health of more than one thousand individually caged monkeys—a daunting task, especially when every single primate engaged in "atypical behaviors."

Carla and I should have been called the psychological well-being duo, not department. Nonetheless, the two of us approached our job as if it really mattered—because it did to the monkeys. And we had our work cut out for us. All the primates deserved and needed special attention, but with the paltry resources afforded our department, only the most severe cases were diagnosed with depression, aggression, hair pulling, feces eating, urine drinking, infant abuse, or stereotypic behavior (circling, pacing, etc.), and only in the most severe cases, self-mutilation. The research industry's literature is pretty clear on the reasons for abnormal behavior: maternal deprivation—taking baby monkeys away from their mothers too early—and keeping socially complex primates alone in small, sterile cages.

For human primates, isolation is considered one of the worst forms of torture. In research labs, isolation is just indifferent efficiency. Over time I noticed that technicians made mistakes in data collection, even without the added variable of a second monkey in a cage. A very small percentage of monkeys were paired in cages; staff found innumerable reasons to keep these lonely individuals separated. The macaque known as 16162 at least had this small bit of fortune—she shared her quarters with another.

I had about fifty cases of macaque self mutilation, some of whom were acutely psychotic to the point where they bit and attacked their own bodies, often doing serious damage. With hundreds of damaged monkeys, and only one staff person working our program—me—I could see that I had been employed to create a paper trail to meet the hollow requirements of the Animal Welfare Act. Clearly neither my position nor my work was of interest to the establishment; they did not care at all whether I actually improved the monkeys' bleak existence. They needed someone on staff assigned to satisfy the weak provisions for psychological well-being, a 1985 amendment to the Animal Welfare Act. This toothless addendum to the act merely requires every lab housing nonhuman primates to have some plan—any plan—for the monkeys' psychological well-being with no minimum requirements or evaluations to prove whether or not the program works. As long as a lab follows their plan, USDA inspectors are powerless to cite, fine, or penalize a research facility for the suffering that monkeys endure under the harsh conditions that are typical there. So I found myself placed in charge of the psychological welfare of hundreds of depressed, frightened, stressed—damaged—primates without the resources to do much of anything to help.

I quickly learned that a behavior diagnosis merely earned a few extra minutes of observation per month, during which time I scribbled notes on a clipboard reflecting whether the primates' condition was improving, the same, or worse. I also employed standard tools of the trade: puzzle feeders, various styles of plastic contraptions where treats could be hidden, grooming boards, a metal plate with a twelve-inch square of stretched fleece (designed to provide some-thing to pull out besides their own hair), and other devices to occupy bored minds.

As one may guess, a Kong dog toy in a cramped, barren cage is a poor substi-tute for matrilineal-family troop interactions (with as many as fifty or more mem-bers) in an ever-changing natural environment. Enrichment devices, however well intended, cannot curb the intractable behavior problems created by captivity in a three-by-three-foot cage. Technicians know this, and often give up filling the puz-zle feeders. We watch the monkeys extract the treats in just a few minutes, often before the Tyvek-clad worker has left the room. Compliance with the behavior program was sketchy at best, partly because it obviously didn't work.

Furthermore, these puzzle feeders and other devices created minor inconve-niences for technicians, who had to fiddle with or move these contraptions to gain access to a monkey. This made me a target of ongoing criticism and ridi-cule since I was the one who kept hanging this "crap" on the cages. Management didn't help any; they also saw me and my job as little more than fodder for a run-ning joke. For example, the operations manager, Art, began his rise to power from humble beginnings as a security guard, yet he entertained himself at every public opportunity by calling me Toy Boy. But as ineffectual and unappreciated as my job was, at least I was able to remain benevolent. I was usually not required to inflict harm, and I did my best to work the program on behalf of these poor, unfortunate monkeys. My heart was breaking for them.

I met 16162 because she had a hair-pulling diagnosis, or to use the sanitized language of the lab, she was "overgrooming." She was a typical Indian-origin rhe-sus macaque with grayish brown fur, a pink, hairless face—pale from her sunless, florescent-lit existence—and a slight build, weighing in at 4.5 kilograms (about ten pounds). But the expression on 16162's face was different. She radiated intelli-gence and curiosity. Her movements were cautious, even thoughtful.

The first time I met 16162, Carla and I entered Room 32 of the Annex Build-ing dressed in matching powder blue lab coats, safety glasses, surgical masks, and latex gloves, then squatted several feet in front of cages A-7 and A-8. Quietly observing the two monkeys in these cages, we tried to ignore the cacophony of other individuals vociferously crashing about in small cages all around us. It ap-peared that 16162 was also trying to ignore the surrounding din as she sat meekly on her haunches at the back of her cage.

"I always make sure she has a mirror," Carla murmured as she moved closer to pick up a scratched metal disk from the concrete floor, wiped off the grime on her pant leg, and hung it by its chain on the outside of 16162's cage. The once-shiny, round surface now barely showed a dull reflection, but after 16162 got used to our presence, she reached timidly between the bars to manipulate the mirror. She held the disk gingerly in her hands, then pulled the mirror halfway into the cage to observe herself with intense concentration, moving the makeshift mirror with slight corrections to broaden her perspective on her dank, concrete-walled surroundings.

"See," Carla pointed out, "she is using the mirror to look down the row of cages!" I remembered watching a movie where an inmate used a mirror to keep an eye out for the guard. This monkey's attempt to catch a glimpse of her captors, or the resident of a neighboring cage, reminded me of the movie. She was trying to expand the horizons of her diminutive world. "She is so smart!" Carla beamed as she self-consciously readjusted her safety glasses and blue paper mask.

Oddly enough, after you've worked for a while in the lab, you learn to read nonverbal communication even with compulsory surgical masks hiding expressions. I suppose we say a lot with our eyes. I could see that Carla took some kind of bizarre pride (or felt misguided personal ownership) in 16162's intelligence. Like Hiram Bingham, who stumbled across the ruins of Machu Picchu, she had discovered a rare individual whom she believed had Einstein-like monkey intellect.

At her core, Carla genuinely cared about the monkeys she tended, and we constantly conspired about ways we might improve their bleak lives. She was emotionally invested, exhausted, and heartsick from years of work in desperate, depressing, and downright cruel conditions. She suffered from sleepless nights, migraines, and what she feared were stomach ulcers, all likely results of her tenure in the lab. I can't imagine how losing such an emotionally crushing job could be a bad thing, but she told me later that it was.

Many people start working with nonhuman primates because they love animals, but when they reach the lab, they either harden their hearts as a survival mechanism—or quit. Quitting seems the better option. There was enormous turnover at OHSU; many people only lasted a couple of weeks—or even a couple of days—on the job.

Not Carla. She had an ambitious five-year plan, which was summarily ignored at weekly meetings with Dr. Scott Murphy, head veterinarian in charge of the Division of Animal Resources. In contrast, I had more modest goals and once asked if the monkeys might receive more fresh fruits and vegetables, noting that this would be good for their psychological well-being. Instead of rejecting the idea outright and risking that he might appear heartless, Dr. Murphy used cold hard science to squash my suggestion and protect the department's budget. He asked that I devise a study to prove scientifically that the monkeys would be better off with more produce. He knew that I was already overwhelmed; designing and seeking funding for an experiment were ridiculously impossible. Not to mention unnecessary because he and I both knew the answer.

The intense whoops and hollers of surprise and joy never failed to demonstrate how much the macaques enjoyed fresh fruit. I blew off his scholarly advice and solicited fruit donations from wholesalers. I also started picking pears, plums, and apples from trees long ago planted on the primate center's 270-acre campus but no longer harvested, I was told, because of liability concerns. I wondered how harvesting fruit could be risky compared with the very real job hazard of contracting herpes B—a deadly simian virus—from a monkey bite.

The truth is that there was never any intention that the psychological well-being program would work. The people in charge kept Carla at the helm of this critically important department while systematically eroding her power and undermining every possibility for success. After I went public after completing my two years of undercover work at OHSU, Carla came under intense scrutiny and eventually lost her job—and her career; she was unable to shake guilt by association from having worked with a spy.

But I'm getting ahead of myself. The day that I met 16162, Carla led me outside to explain why she always gave the macaque a mirror: "Once, when I was observing 16162 while introducing her to a new cage partner, she had her back turned to the other monkey in an appropriate display of submission but was cleverly using her mirror to keep a close eye on the stranger in her cage." Carla's initial motivation to give 16162 a mirror was based on an earlier observation of her behavior noted several years before: "previously showed interest in the cage flap on her cage, seemed to use it as a mirror, (would hold it up at middle of eyes as if trying to see above and below)." Some of the older cages had a flap, a small, swinging piece of metal attached above the slot where technicians shoved food into the cages, presumably to protect human hands from monkey scratches. Obviously, 16162 had figured out her own use for this flap.

Scientists have used mirrors for decades in attempts to unlock the mysteries of self-awareness in nonhuman primates. Do apes and monkeys have self-consciousness—awareness of their individuality? The gold-standard test, developed by Gordon Gallup, is the mirror test: a mark or colored dot is placed on a primate's forehead to see if he or she will react when looking in the mirror by using a hand to find the dot. Of the primate species, only great apes, human beings, chimpanzees, orangutans, and possibly gorillas have conclusively examined their faces in a mirror, noticed the mark, and then touched their foreheads. These species clearly recognize that they are looking at themselves in the mirror. Until recently, monkeys had not passed this test. Instead, they reacted to their face in the mirror as if it were another monkey, possibly a rival. But in 2009, comparative psychologist David Smith cited "growing evidence that animals share functional parallels with humans' conscious meta-cognition [and]...that some animals have functional analogs to human consciousness" (Smith 2009, 389). I could see that Carla dreamed of being the one to make this discovery, but 16162 was killed before she had a chance to demonstrate her self-awareness.

We continue to learn about similarities that connect human beings with other species. We share twenty-three million years of evolution with great apes; we diverged into a separate species just six million years ago. Duke University researchers have demonstrated that monkeys have the ability to perform mental addition. In fact, monkeys performed about as well as college students given the same test. Ironically, more than fifty years of such research led Thomas M. Burbacher and Kimberly S. Grant to coldly conclude that monkeys were great subjects for research on the functional effects of exposure to neurotoxic agents because the "behavioral repertoire of nonhuman primates is highly evolved and includes advanced problem-solving capabilities, complex social relationships, and sensory acuity equal or superior to humans" (Burbacher and Grant 2000, 475).

It seems to me that discovering emotional and intellectual similarities across species is only important if it leads us to reexamine the ethics of exploiting nonhuman primates. Unfortunately, ethics are difficult to find in a research setting, even when empathy is the subject of a particular study; scientists always seem to find a way to add a sadistic twist. For example, Jules Masserman and his colleagues found that monkeys whom they had trained to pull two chains to gain different amounts of food usually did not pull the one that offered the larger reward if doing that caused pain to another monkey. After the macaques witnessed the effects of a shock on another monkey, two-thirds chose the nonshock food chain even though it offered only half as much food. One of the macaques stopped pulling the chains altogether for five days, and another for twelve days, after seeing the effects of the shock on another monkey. These monkeys literally starved themselves to avoid harming another monkey (Masserman, Wechkin, and Terris 1964, 584). What is most "shocking" is that researchers never directed their attention within to examine their lack of empathy in light of their findings.

In my experience, once research funding runs out, scientists always come to the same conclusion: their findings are of great interest to humanity; however, more research is necessary. Judging from all the Saabs, BMWs, and Mercedes in the parking lots of the facility where I worked, animal researchers' motivations may have more to do with making the next car payment than any lofty humanitarian or scientific goals.

With my overburdened workload, I could only visit 16162 about once a month to assess and try to mitigate her overgrooming problem. Unfortunately, but not surprisingly, things went from bad to worse. I tried everything, including soothing ocean music combined with cardboard tubes and cherry wood for her to chew on. However, 16162 banged her puzzle feeder roughly against the side of the cage to avoid the hassle of removing treats one at a time, effectively outsmarting the device and defeating its purpose. She plucked the hair from most of her body as efficiently as she pulled the fleece from her grooming board. But I made sure that she always had a mirror, and she used it proficiently. Once I even stole enough time away—by skipping a break—to videotape her using the mirror. It was clear to me then—as it is now—that she knew exactly who was looking back when she gazed into the mirror.

One day, when I went to check on her, I found 16162's cage empty. I looked her up in the computer and found that she had finally been moved to the clinic. I rushed to the large window that separated the busy hall from the row of ill monkeys in the clinic. It made no sense to subject sick animals to the constant stress of being on display behind glass with lab workers passing all day long, but that was the way the clinic was set up.

It was easy to spot 16162 in the row of cages hanging on the wall, crouching on a tiny metal perch in the back of her cage. She looked horrible. She had a history of diarrhea and was so thin that I had already added a diagnosis of anorexia to her behavior description. She was too sick for treats. There was nothing I could do for her.

She spent the next two months fading away in the clinic. When I walked past the windows of the hospital room, it was impossible not to notice her. She looked pathetic. Crouching in the corner of her cage, she held her hands over her head as if trying to hide.

Carla asked if 16162 could be given pain relief. The clinic veterinarian, Dr. Gloria McDonnell, glanced at the monkey and replied, "She's just feeling sorry for herself." Dan, another technician standing nearby, quipped, "Yeah, is there a computer code for pouting?"

Gloria sedated 16162 with Ketamine and ran a duodenoscopy to try to find out why she was so sick. The notes of this procedure read as follows:

> Positioning was in left lateral recumbency. An oral speculum was placed, and the 8.5 mm endoscope was inserted orally and advanced through the stomach to the distal duodenum. A series of 7 mucosal biopsies were taken in the distal and mid duodenum. No obvious lesions were noted. The insufflation was reduced and the scope withdrawn. Recovery was uneventful. (ONPRC medical records retrieved from Oregon public records as documented by In Defense of Animals.)

The tests from this procedure came back, but there was no medical explanation for 16162's condition.

The primate center at OHSU called 16162's condition chronic or stress-related diarrhea. This condition was common among lab primates, even for rhesus macaques, a species known for strong constitutions and consequently often used in research. These macaques are known for their ability to survive years of duress in a cage; they are sometimes called a "weed" species because of their resilience. But research takes its toll on every animal. Beyond the stress of experimentation, lab monkeys are constantly moved, and these highly social animals never have a sense of community. They suffer from loneliness and the constant making and breaking of bonds. What few monkey pairs exist seldom persist for any length of time. Additionally, they are frightened when their cages are hosed down, which happens daily with the macaques inside.

Finally, 16162 was put on "the schedule," and when her time came up—exactly one week later—she was killed. Necropsy, the animal equivalent of an autopsy, ended her life—in animal labs, necropsy is the way to kill an individual. Those who perform necropsy are trained to cut internal organs and tissues while the animals are still alive. This ensures that the organs are infused with blood. Dissected parts and organs are placed in plastic bags, boxed, and put on ice to be delivered to a scientist—all done with casual nonchalance.

Poor 16162 was groggy from the Ketamine injection when she arrived for necropsy, schlepped in a Goi box (named after its inventor). A Goi is a metal box used to transport monkeys safely from one place to another. She was laid out on her back on the cold stainless-steel table, further sedated, then opened up with a scalpel. The technician, skilled from "sacrificing" hundreds of other monkeys, made short work of 16162's procedure. He systematically removed both ovaries, mesenteric lymph nodes, thyroid, pancreas, urinary bladder, and sciatic nerve, as well as other minor body parts, then took a massive syringe and continued to draw blood from a major artery near 16162's exposed heart until it finally stopped beating. After almost nine years of life—each day filled with boredom, sickness, and depression—16162 was unceremoniously parted out for further exploitation.

Her human captors never loved 16162. In writing about her life, I realize that even I wasn't able to love her in a way that was meaningful to her. I never had the time to really get to know any of the monkeys at OHSU. Even ones like 16162, who made an impression on me, were denied their most basic physical and emotional needs. What would happen if research employees like me came to know the primates living in those stainless-steel cages?

As taxpayers, we all contribute to animal experimentation, but we are kept in the dark about what actually happens in these facilities. It took eight years—and a lawsuit orchestrated by In Defense of Animals—to force OHSU's primate center to produce public records on the monkeys.

It's hard to wrap my head (and heart) around primate—or any animal—experimentation. Researchers and staff exploit and seriously damage sentient beings, yet refuse to satisfy these animals' most basic needs. It was stress that killed 16162; she communicated her desperation by pulling out her hair and ultimately becoming very ill. She was just one among tens of thousands of monkeys at the Oregon Regional Primate Research Center and around the country and the world. And monkeys are just one species among many millions—including rabbits, mice, guinea pigs, rats, pigeons, pigs, dogs, cats, cows, chimpanzees—that we exploit in the name of science every year. The day cannot come quickly enough when we abolish animal testing.

While she was in the clinic, Carla and I gave 16162 her mirror, but she never used it again. The last spark of life had gone out of 16162 long before they killed her. The last memory I have of 16162 is her crouching in the back of her cage with dark, sunken eyes. Just a few clumps of fur stuck out here and there on her emaciated body, some of which made a halo of disheveled hair around her bald head.

I wasn't sad when 16162 was killed—for a monkey in a research lab, life is worse than death—but I will always regret that I did not even give her the small dignity of a name.

8

Monkeys, Malaria, and My Work in Miami

Juan Pablo Perea-Rodriguez

As a child, I traveled through Colombia with my family, marveling at the beauty and diversity of fauna and flora. I was intrigued by the beautiful waters and mountains of the Cocha lagoon in southern Colombia near the Ecuadorian border, the gigantic cordilleras that extend throughout most of the country, and the blue-green waters of La Guajira. As I became more familiar with the natural world, I developed an interest in biology—I wanted to understand nature. One month before I turned eighteen, and immediately after I graduated from high school, we had to leave Colombia because of the growing political crisis. In 2001 we moved to the United States and settled in Miami, Florida. Still fascinated by nature, I decided to pursue a degree in biology to improve my understanding of the natural world in the hopes of one day adding to worldwide conservation efforts and perhaps even contributing to saving my native Colombia.

As chance would have it, a student internship at the DuMond Conservancy for Primates and Tropical Forests in Miami changed my life. The DuMond Conservancy is a scientific organization that provides spacious, outdoor homes to nonhuman primates, most of whom have been retired from biomedical laboratories. DuMond also helps educate the public about primates and their environments through noninvasive research. At DuMond the best-represented primate is the owl (or night) monkey. Other species at DuMond Conservancy, although native to Colombia, were completely new to me.

Owl monkeys are native to Colombia, where they are known by the locals as martejas. They can be found in the rain forests of Panama clear down to the sparse grasslands and forests of northern Argentina known as the Gran Chaco. Owl monkeys are the world's only nocturnal monkeys. Their scientific name, *Aotus*, means "without ears," which is a bit misleading. Although not very noticeable, their ears are quite large, and they are highly sensitive to sound. Due to their nocturnal habits, these primates have large, shiny eyes and an acute sense of smell, through which they investigate their surroundings and one another. Owl monkeys are omnivores, having long, thin fingers with which they capture insects.

These monkeys live in small families, which include a reproductive pair and up to three offspring. One of the must puzzling aspects of owl monkey behavior is their mating system. They are one of just a few species of monogamous primates. Although gibbons, siamangs, some prosimians, and a few other New World species share this characteristic, monogamy is rare in nature, especially among mammals. Owl monkeys bear one offspring per year, tended largely by the father for the first couple of years.

Owl monkeys have white faces with distinctive black marks on their foreheads and eyebrows, marks that are unique to each individual. Their brownish and grayish fur grows darker farther from the head, culminating in a long, black, slender tail. The skin on their faces varies from pink to black. Their dark hands are covered with wispy, whitish hairs. Owl monkeys are separated into two distinct groups, distinguished by the color of their throat and neck. Northern owl monkeys, such as the Colombian and Panamanian species, have a grayish hue and are therefore considered to be part of the gray-necked species. The monkeys that live south and east of Colombia all the way down to Argentina have a distinct coppery hue on their necks and are therefore called the red-necked species. Owl monkeys living in cooler latitudes have fluffier fur. The Argentinean-Paraguayan monkey (the southernmost species) experiences the harshest winters; a thick coat disguises how small they are underneath their opulent hair.

When I started my internship at DuMond, I cared for owl monkeys, recorded data on their natural behavior, and provided enrichment. During hurricane season, many of us supplied on-call assistance. Caring for more than sixty owl monkeys in a hurricane is a challenging task. More routinely, I and other interns added foliage, such as flowers and fresh clippings from native trees, to enrich the primate's enclosures. The primates clearly enjoyed eating the fresh greens that we provided.

Students at DuMond help out with scientific studies on foraging behavior, parental care, and activity patterns. We also recorded sexual behavior and the development of offspring. We were especially interested in the way these primates interacted with each other when they were choosing a mate. As part of our behavioral studies, we added enrichment bottles (filled with their favorite treats, in most cases apples and grapes) to their areas and recorded their interactions. These studies, designed by Dr. Christy Wolovich to explore the role of food sharing in paired groups, both challenged the monkeys' intellects and rewarded them with tasty treats.

Other studies required us to stay up all night to observe nocturnal behavior. We knew that owl monkeys are lunarphilic, or "moon lovers," that are most active on full-moon nights. We studied and recorded changes in their behavior at different times in the lunar cycle to try and learn how their behavior varies in high or low moonlight. Toward this end, we asked questions, for example, do owl monkeys socialize more when the moon is full? Do they forage more with increased moonlight? During these intense night observations, I was gifted with a rare sight—the birth of an owl monkey. Just hours before sunrise, Ophelia was born, and I witnessed rare social behavior, such as grooming and mating.

These monkeys have spent most of their lives in small cages in laboratory settings. Many are elderly and thus require extra care, including special diets and daily medication. Some were caught in the wild, then transported to labs for experimentation. The colony where I worked included owl monkeys that came from labs engaged in surgical procedures on eyes. Due to their very large eyes, which enable them to absorb more light at night, biomedical researchers consider owl monkeys valuable tools and exploit them, performing surgeries that cause visual impairment and, in some cases, complete blindness. Some of the primates who arrive at DuMond are missing eyes. Many of the owl monkeys have noticeable milky scars on their eyes from these intrusive and destructive procedures.

Like many people who work rehabilitating primates, I was shocked to learn how many animals are used in biomedical laboratories. Dumond continues to receive retired owl monkeys from the CDC (Centers for Disease Control), where they are used in herpes and malaria research. In these tests, innocent primates are injected with diseases so they can be used to test pilot drugs. The researchers' goal is to develop medicines that can treat disease or prevent epidemics. While these researchers have the worthy goal of helping humans, their means are immoral. Thousands of primates are selfishly sacrificed in the hope of some gain for humanity. I am sure that these researchers would not like to be exploited this way.

When I learned that these owl monkeys were being used for malaria research, I was reminded of the story of a young Colombian pathologist, Manuel Elkin Patarroyo. Most of us who have lived in Colombia learned that he had cured malaria back in the 1990s. This doctor was known in Colombia for inventing the first synthetic malaria vaccine. He was much respected because we had heard that he had given the rights to the vaccine away free of cost. Little did we know that his vaccine was ineffective, and we were unaware of his atrocities: cruelly exploiting owl monkeys.

In 1994 Patarroyo claimed to have developed the first synthetic malaria vaccine: Spf66. Because this drug was synthetic, it was very cheap to manufacture and therefore accessible even to the poor. Consequently, Patarroyo's new vaccine received much attention. He followed up his preliminary studies with human trials in Colombia, Thailand, and Tanzania. He claimed a 40 percent success rate, which, although not perfect, would save millions of lives. Unfortunately, when several international organizations tried to duplicate his study, results showed that those who were treated with Spf66 and those who received a control substance experienced the same effects.

Patarroyo offered many excuses to explain why these studies did not discredit the effectiveness of his medicine. People tended to believe him, at least for a decade, because Patarroyo had gained celebrity status. Nations such as Spain and the United States continued to send money to fund his research. The hope of a Colombian cure for malaria kept on growing. Only in the late 1990s did Patarroyo's reputation become murky when journalists began to uncover some of the shocking truths about his research, including what went on in his lab in the Amazon. He was accused of trafficking owl monkeys across the Colombian/Brazilian and the Colombian/Peruvian borders. Journalists also exposed the inhumane conditions at Patarroyo's laboratory, including monkeys who died of malaria. Patarroyo began to look very much like a mad scientist who had long been hiding his gruesome work deep in the thick jungles of the Amazon.

Malaria takes hundreds of millions of lives a year, disproportionately affecting pregnant women and children. That is why most people hope that someone will produce an effective vaccine soon. Also like most people, I am opposed to the inhumane treatment of animals and their exploitation in biomedical research. Patarroyo was notorious for his cruel treatment of primates. He claimed that monkeys in his facilities lived in "hotels" where they checked in for three months and were then returned to the exact tree where they had been captured, unaffected by his experiments, totally unharmed. Of course, this was a lie.

The Colombian government investigated Patarroyo's laboratory in the Amazon and found that many monkeys checked in but didn't check out. They also found many sick and dying monkeys, and they discovered nonnative owl monkey species, indicating that he had trafficked Peruvian and Brazilian primates without permission from these neighboring nations. By promoting the trafficking of monkeys, Patarroyo created incentive for villagers to hunt these endangered species. Poachers earn roughly fifty dollars for each monkey they catch, which is a great deal of money in Colombia. (These monkeys cost three thousand dollars in the U.S.) Many people wonder why Patarroyo has never been imprisoned since the natives who captured the monkeys surely would have been if they had been caught.

Patarroyo's laboratory lacked a rehabilitation plan for his test subjects. If released, these exploited primates faced La Isla de los Monos—Monkey Island—a small island in the middle of the Amazon River inhabited by plenty of wildlife and a small population of humans. Nonetheless, Patarroyo released his test subjects—once he had no more use for them—onto this island, perhaps imagining that the monkeys would somehow recuperate from the deadly diseases he had given them, and the terrible ordeal he had put them through, and live in harmony.

Either Patarroyo has very little knowledge of the behavior and ecology of owl monkeys, or he simply does not care about the animals he has exploited. Owl monkeys are territorial. While studying owl monkeys in Argentina, I witnessed a fight between neighboring social groups that left one monkey with no tail. In spite of the potential for violence in cramped territories, Patarroyo admitted to releasing more than twenty-five hundred monkeys onto this one-thousand-acre island in the last thirteen years. In the wild, owl monkeys' social groups (three to five monkeys) claim four acres. For these monkeys to have enough territory, the island would have to be twenty-five hundred acres. Patarroyo has put two and a half times as many owl monkeys on Monkey Island as this area can comfortably sustain. Furthermore, among monogamous species, populations are generally composed equally of females and males. Did Patarroyo pay attention to how many males and how many females he released onto this small island? Did Patarroyo ponder the damage he would cause to individuals, communities, and the environment? How would diseased monkeys affect the human inhabitants of Monkey Island?

Kidnapping this endangered primate from Colombian, Peruvian, and Brazilian forests for decades of Patarroyo's scandalous experiments has caused considerable ecological damage. Local and neighboring villagers have targeted owl monkeys as an easy source of income, further reducing the numbers of this endangered species. Trees were felled so that Patarroyo's hunters could access these monkeys in their homes and steal them away to his laboratory. Because Patarroyo has imported owl monkeys from neighboring lands and released these nonnatives into Colombian territory, he has altered the distribution of species and enhanced competition for food, mates, and territory for Colombian owl monkeys. This is likely to create hybridization among neighboring species. It is difficult to tell what the long-term effects of Patarroyo's biologically careless, self-indulgent behavior will be.

In contrast to Patarroyo, the director of the DuMond Conservancy, Dr. Sian Evans, works for monkeys (instead of exploiting them for her purpose). She has studied owl monkey behavior for twenty years. She cares for and knows each resident, and in her late-night rounds through the "owl monkey woods," she observes each individual carefully, inspecting his or her overall condition. Although it would be best if these monkeys could live wild in their native lands, they are very lucky to be under Dr. Evans's care. At the DuMond Conservancy, monkeys enjoy Miami's subtropical weather, including natural sun and moonlight. In their outdoor enclosures, located in a natural hammock forest, they forage for insects and small vertebrates and eat flowers and fruit in the trees that surround their houses. At night, when they are most active, they call to each other with their distinctive hoots. Their home away from home is designed to reduce stress and provide enrichment for these unjustly exploited primates.

At DuMond we try to compensate for the unfortunate experimentation these monkeys have suffered. We also try to prevent further primate exploitation.

For example, we educate people about owl monkeys, sharing our fascination with these marvelous beings and explaining to visitors how much about their behavior and evolution is still a mystery, perhaps because they are nocturnal. We encourage young scientists to consider studying owl monkeys in the jungles of Colombia, working to help these delightful creatures, rather than exploiting them.

Some people, like Dr. Evans at DuMond, sacrifice much of their lives to protect the well-being of wild and captive primates. Other people, usually influenced by greed and selfishness—or simply misguided—undermine these conservation efforts. Patarroyo probably believed for many years that he was close to producing a miracle vaccine that would save humanity from malaria. No doubt he was also caught up in his personal fame and glory, fearful of failure and the loss of this elite status, and so he continued to waste government funds while trashing Colombian fauna and flora in pursuit of a fantasy drug that never materialized. But the destruction he caused—both to individuals and ecosystems—was all too real.

If nonhuman primates could defend themselves and their homes, humans would not need to do so. But they cannot protect themselves from human predators, so we must. Those of us who know and care about primates—or any animal who is exploited or endangered—must make it our life's mission to defend and protect these individuals and species. It is up to us. I am working to protect Florida's Everglades. What will you do?

Acknowledgment

Thanks to the editor, Lisa Kemmerer, for helping to create an essay with polished English.

9

Learning from Macaques

Linda D. Wolfe

The other animals humans eat, use in science, hunt, trap, and exploit
in a variety of ways, have a life of their own that is of importance to
them apart from their utility to us. They are not only *in* the world,
they are *aware* of it. What happens to them *matters* to them. Each
has a life that fares better or worse for the one whose life it is.

Tom Regan, "The Philosophy of Animal Rights" (italics added)

Through field research, I came to understand that other primates (primates other
than human beings) are sentient and conscious. Since 1972 I have observed rhesus (*Macaca mulatta*) and Japanese macaques (*Macaca fuscata*) and, to a lesser
extent, Balinese macaques (*Macaca fascicularis*) and a prosimian species known
as lemurs (*Lemur catta*). Macaques are Old World monkeys of the order of Primates. Once I came to appreciate the mental capacities of nonhuman primates, I
became aware of other animals generally and viewed all creatures in a new light.

Over the last thirty years, I have studied free-ranging Japanese and rhesus
macaques in several settings, including India and Florida. Needless to say, rhesus
monkeys are not native to Florida, but a troop was transplanted to this suitable
climate in 1938. I also studied monkeys on a mountain near Kyoto, Japan, and on
a ranch north of Laredo, Texas. Again Japanese monkeys are not native to Texas,
but a semifree ranging troop was relocated there in the early 1970s.

Caretaking

As I observed these macaques, they revealed their intelligence and sensitivity, showing me that they have thoughts and emotions very much like my own. I also found strong bonds of affection and diligent caretaking in macaque troops, much like human behavior.

Occasionally normal troop relationships broke down, and a fight ensued. Following a fight, there was reconciliation and the monkeys groomed each other to reestablish normal social relations. Larger fights, however, often resulted in wounding and blood loss. Human beings are not the only primates with tendencies toward violence.

Nor are we the only primates who tend our wounded. In 1973, while studying a semi-free-ranging transported troop of Japanese macaques, I observed a fight where the second-ranking male was badly injured. Following the fight, even the aggressors huddled with the injured male and groomed his wounds for twenty-four hours. During this time, the injured male was never without the body warmth of several monkeys, who sat close enough to touch him. Grooming and huddling with their injured cohort suggest that monkeys somehow recognize that wounds need to be kept clean and an injured body should be kept warm. In every case of injury that I observed, grooming and huddling followed, which no doubt saved the life of the injured monkeys.

In 1977 in Arashiyama, Japan, when I was studying the behavior of another troop of macaques, I again noted a mother's caretaking while watching a young mother with her first offspring, whom I called Hajime. Hajime's mother was an adequate caretaker, as young mothers go, but she let her son wander away at an early age. Consequently, when Hajime was a little over a year old, he was injured, perhaps by falling out of a tree. He appeared to have two broken arms or perhaps broken shoulder blades. He could not walk or put pressure on his arms and could barely even raise an arm to eat. In his time of need, his mother started to care for him again. He rode awkwardly on her back, holding on with one hand. Hajime's mother protected him from his playmates and made sure he got enough to eat, though he ate slowly. Eventually he healed, thanks to the special care of his mother.

I was riding in a three-wheeler taxi, traveling along a wide four-lane highway known as Agra Road, when I observed another incident illustrating a mother macaque's strong attachment to her offspring. It was 1988, and I had the opportunity to study rhesus monkeys, a species of macaques native to India and closely related to Japanese macaques. This study took place in and around Jaipur, India, where I learned that a troop of rhesus monkeys lived on either side of Agra Road and sometimes crossed this busy highway. As we traveled along, an adult female monkey, followed by a two-year-old, ran across the highway in front of us toward the rest of their troop. The juvenile was sending out distress calls as he tried to keep up with the female—presumably his mother—who was running just ahead of him. Cars were slowing down to let the monkeys pass, but a large truck hit and killed the young monkey.

When the female reached the other side, she looked back and saw her child lying motionless in the road. As she looked at her dead offspring, she touched her abdomen, using the gesture that mother macaques employ to tell their young to climb under their mother's abdomen, where they are carried. She started back across the highway toward her dead progeny, but the troop screamed in apparent distress. Obedient to their call, she turned and, after one more glance backward, joined her troop and disappeared into the Sisodia Rani Ka Bagh gardens. Had the troop not prevented her, it appeared that she would have run back into the street to her child's body and also been killed.

Because macaques live in many Indian cities, autos are often deadly for this threatened species. In addition to rhesus macaques, dogs, goats, and cows also roam the streets of Indian cities. Dogs, of course, are the main predators of macaques throughout India and the rest of Asia. One morning in the city of Jaipur, I was observing a group of monkeys who were enjoying food fed to them by people headed for the local Hindu temple. When a couple of dogs appeared, the monkeys took off up the sides of the buildings. One old female slowly climbed a drainage pipe with a newborn on her abdomen and a one-year-old on her back. She clearly recognized the danger the dogs presented, yet she carried both of her burdensome progeny, which slowed down her escape considerably. She seemed determined to rescue her offspring from the dogs. Struggling mightily, she managed to carry both young ones to safety. Given her older age and the extra weight of a yearling, she barely made it before the dogs reached the wall. It would have been easier for her to climb to safety alone, but if she had, it is possible that neither of her little ones would have been able to move fast enough to escape the dogs. She risked her life for the safety of her young.

Lessons Learned

Studying macaques taught me much about these nonhuman primates, much about myself, and much about animals in general. Through fieldwork with macaques, I became a vegetarian—I quit eating the flesh of any creature. During the year I spent in Japan in the mid-1970s, I ate very little beef or chicken. When I returned to the United States, I noticed that it took hours for me to digest beef and during that time, I felt like a sloth. Chicken no longer tasted good, so I stopped eating both chicken and beef, which put me on the path to vegetarianism.

A few years later, while studying rhesus macaques in Jaipur, India, where cattle freely roamed the streets, I discovered that cows also have personalities and intelligence. Needless to say, they also suffer and prefer to remain alive, rather than be consumed. This insight confirmed my decision to stop eating animal flesh.

The Next Generation

As a university professor, I am in contact with many undergraduates in liberal arts core courses, as well as graduate students in advanced courses in biological anthropology, including primatology. As part of the curriculum, I discuss the ethical treatment of primates, helping students become aware of their physical and psychological needs.

Unfortunately, students too often assume that whatever procedure a researcher deems necessary ought to be done with little or no concern about the pain and suffering of the animals who are exploited. I bring to the classroom the idea that pain and suffering are always important. I suggest that students should be open to more enlightened ways of understanding and therefore avoid causing pain and suffering to nonhuman primates. As individuals of conscience, professors ought to encourage students in biomedical research to pursue careers that use modern alternatives to vivisection (the exploitation of live animals for experimental purposes).

Graduate students—MA and PhD students—can choose to study under the tutelage of a particular professor. Therefore, graduate students usually find a professor with interests similar to their own, someone likely to provide the desired research experience. The relationship between a graduate student and a professor is somewhere between that of colleagues and a student/teacher one. Ideally professors of primatology discuss with their graduate students not only primate evolution, adaptation, and behavior but also ethical issues that stem from exploiting primates for biomedical research, the importance of environmental enrichment for those unfortunate primates, the critical problem of habitat destruction, legal and illegal international trade in primates, and the very real possibility of extinction that may result. Through such discussions, the professor has the opportunity to influence the next generation of primatologists and help them understand that nonhuman primates are conscious and sentient beings whose interest in living a natural life ought to be honored, and that we should do all we can to prevent their extinction.

Most humans now recognize that primates suffer when they are caged for biomedical research or confined as pets to entertain humans. It should be illegal to own or buy and sell primates as pets or display these individuals in roadside zoos or pet shops because they suffer a great deal from such exploitation. The psychological damage caused by isolation and experimentation is intense and lasts throughout a primate's lifetime.

In years of research and teaching, I have noticed that several topics relating to the exploitation of primates in biomedical research are missing from the ongoing dialogue, including the following:

1. Why is human suffering considered a greater evil than pain purposely inflicted on primates in experimentation?

2. What formula might be established to measure suffering and calculate how much misery is justified in exchange for hoped-for human benefit?

3. While euthanizing a psychologically and/or physically damaged primate covers up the evidence, removes the problem, and is less expensive than maintaining sanctuaries, should we not pay the cost of retirement for these primates as partial compensation for their suffering due to exploitation and imprisonment?

4. At what point should a primate used for experimentation and testing be retired, either to a no-kill sanctuary or through euthanasia? How much torment may we inflict, and for how long, on one helpless individual in a laboratory?

Discussing these issues in the classroom, as well as in labs, sanctuaries, and the community at large, will further the development of alternatives to vivisection, reduce the use of primates in biomedical research, and lead to more humane relations between human beings and other primates.

10

The Winding Path to Where I Stand

Becoming a Primatologist

Debra Durham

The Physicians Committee for Responsible Medicine (PCRM) promotes higher standards of ethics and effectiveness and alternatives to the use of animals in research, education, and training. Research and testing cause pain and suffering and provide unreliable information regarding human health and welfare. I work as a senior research scientist with PCRM, advancing the use of nonanimal research methods and testing and promoting higher ethical standards. Nearly all of my work revolves around vivisection. Vivisection is experimentation on living beings and includes experiments performed in private and government labs, testing products and chemicals mandated by the government, and voluntary testing of products like cosmetics, foods, and beverages by corporations large and small.

Generally speaking, the U.S. government does not track individual nonhuman primates used in research and testing. Thus, it is exceedingly rare and difficult to know what happens to specific individuals who are exploited for vivisection.

I mention individuals because, for me, the point of reference is the individual, and this view is central to my ethical position and advocacy. That said, I rarely know the individuals for whom I advocate—their names or their faces. I don't need to. I stand up for them just the same. The capacity for empathy and care allows me to consider the experiences, desires, and needs of every individual. This is the basis for my concern over discomfort, suffering, and physical and psychological pain in others. I don't need to know an individual to be aware of his or her capacity for these experiences, and so I advocate for the nameless and countless nonhumans exploited by science and the cosmetics industry. A legal complaint, a sidewalk demonstration, or a letter to a member of Congress, for example, all help expose the abuse of nonhumans.

Foundations and Transformations

Some of my earliest experiences with monkeys took place in laboratories. The fascination that first drew me to the lab stemmed from studies in biological anthropology. I wanted to see and work with primates. However, excitement at the prospect of working with primates turned into an unwanted and increasingly unbearable immersion in the sadness, despair, pain, and suffering of the everyday lives of these monkeys. Initially I worked in an infant primate research laboratory, where baby baboons and long-tailed and pig-tailed macaques were kept who had either been rejected by or intentionally taken from their mothers for research. Newborns were kept in incubators in a nursery like human infants in a maternity ward only they were kept alive to be poisoned, tortured, and killed, not nursed to health and returned to their mothers' arms. How could any well-adjusted human fail to feel sadness, despair, and pain in such a facility?

When the little monkeys were a few weeks old and could physically move and eat on their own, they were moved to another room, where they were kept for the next year or so. The room consisted solely of stacks and stacks of orphans in cages. These orphans were not kept together; rather, each was locked in his or her own prison, lined up side by side, one on top of the other. Motherless monkeys each had what was called a surrogate—as in surrogate mother. They were not the spiked wire frames with flashing lights for eyes that Harry Harlow used in his infamous experiments, but these surrogates were still no substitute for a mother. They consisted of a piece of PVC pipe covered in diaper flannel, hanging from the ceiling of the cage by a chain. The suspended tube moved a little and was therefore presumed to be more natural. Still, these needy, deprived babies clung to these cloaked PVC pipes, sucking their thumbs.

I remember thinking it was cute at first, then slowly becoming aware of how sad and pathetic these experiments were. The babies wanted a mother— they needed a mother—a warm body. They needed to be cradled and nursed and groomed and spoken to in the grunts and puffs and lip-smacks of monkey-speak.

Some of these wee orphans lay in fetal positions on the cage floors, clinging to a diaper or clutching themselves as if they might somehow duplicate the comfort of a mother by wrapping their arms around their own little bodies. Sometimes they screamed in despair and thrashed around. Other times they just lay there, sometimes looking around at the room and the other monkeys or lost in their own thoughts, some kind of mental auto pilot—barely responsive.

I doubt anyone is surprised to learn that babies who are deprived of their mothers suffer behavioral, physical, and psychological pathologies that become worse over time and plague them throughout their lives. Likewise, mothers suffer acutely when their babies are taken away from them. Although vivisectors and government officials are aware of these (and many other) tragic effects of human experimentation on nonhumans, they fail to acknowledge the ethical issues that are inherent in any experiment that purposefully causes and enhances maternal deprivation or any other social trauma. Some even consider such experiments humane, citing government funding or noting that such tests are licensed and legal, as if paperwork has the power to purify.

Some babies cowered in the backs of their cages; others fought and screamed and did anything in their power to keep me from taking them out of their cages. My job was to put them on a scale for a moment or measure the size of one of their feet (with something that looks like the tool used to measure feet in a shoe store). I didn't intentionally hurt them, but the babies were terrified all the same.

These desperate orphans often responded to me—especially if I had food or pretended to have some. Certain individuals stand out in my memory nearly fifteen years after the fact. For example, I remember a monkey whose shorthand identification (an ear tag consisting of tattooed letters) was AB. He was a very handsome pig-tailed macaque—at least to me. He had been reared by his mother in a group—albeit a captive group—before he was brought to the lab. When I first met him, he was curious and spunky, but his demeanor changed quickly. He soon became withdrawn and fearful.

I was sad for him, and I felt bad to make his day of isolation in a cold steel cage even worse by handling him. I wasn't sure how to ease his pain. I held him close, spoke softly in his ear, and lip-smacked for him. When we went for X-rays (about once per month), I let him out of the transfer box—a plastic lunch cooler with holes drilled in the top—so that he could explore the exam table. I held him, tried to play with him, and gave him extra time outside of his tiny cage. I let him cling to my lab coat, belly to belly, the way he must have held to his mother under better circumstances. One day when I was holding him, preparing the table, he buried his head in my shoulder. It broke my heart. I had all the power, and he had none. He wanted to hide. He wanted protection. I couldn't protect him—or at least I didn't.

AB began a transition for me. I knew I couldn't continue to work with these babies. Their suffering troubled me; I didn't want to hurt them. I didn't like what was happening to them, but I could not yet articulate my moral discomfort with what I was seeing and doing in those labs. I was still far from finding my voice and the courage to speak out against primate experiments. AB was a turning point all the same.

First, I sought a different role within the same facility, which turned out to be not so different after all. I took a job working for the benefit of the animals—environmental enrichment—but there was still plenty of suffering, especially among those who had been held captive for a long time. The primates were still exploited unjustly.

At some point, I heard that AB was no longer in the infant lab. I didn't know if he had died or been moved. I was sad, but I didn't really want to know where he was or why. It was easier if I didn't ask, if I didn't care.

In 2006 I finally realized that I wanted to know the truth about AB. While working for People for the Ethical Treatment of Animals (PETA), I began investigating the lab where I had first met AB. I reviewed records from the primate center, which I requested under state laws concerning government documents, sometimes called "sunshine laws." I requested veterinary records for AB, but a note informed me that his files had been destroyed some years earlier. It turned out that AB had been killed shortly after he was moved. I cried, mourning AB's short, tragic life and completely unnecessary death.

A few weeks after I learned that AB had been killed, I attended a candle-light vigil for primates in laboratories and held a sign with a photo of a baby monkey that read "In honor of J95350," which was AB's identification number. And when I presented a paper about primates at a scientific conference last year, I listed AB in the acknowledgments. I've told his story in media interviews, to students at campus presentations, and to other activists I work with. AB was in a windowless room in a windowless basement in a cage where few people even knew whether he lived or died. But I knew, and knowing affected me in ways I did not anticipate. I tell my story of knowing AB to honor his life and death and expose his suffering. Though he is gone and the time when I might have helped him has passed, AB had a tremendous impact on my life and my work in helping other nonhuman victims of science.

Flash Forward to Advocacy

From 1999 through 2001, I was a doctoral research student in southeastern Madagascar examining two lemur species: the red-bellied and the red-fronted brown lemur. I followed lemur families that lived in heavily logged or unlogged forest, monitoring where they went, what they did, what they ate, and whether they had babies. By comparing results from these different sites, I noted that logging affected lemurs—the way they used time and space, what they ate, and their reproductive patterns. While my early experiences in the lab drew me to advocacy, it was in Madagascar that I first focused professionally on individuals and on protecting them.

Because my background is in animal behavior with a specialization in primates, my work with PCRM focuses on analyzing experiments performed on primates. I also review information about primate vivisection to uncover and expose violations of laws and regulations, and common decency. I spend much of my time studying the psychological symptoms and disorders experienced by primates, including chimpanzees, who have been exploited for laboratory research and now live in sanctuaries. This feels very distant from my work in the jungle of Madagascar, studying wild primates and working on conservation issues—which is what I thought my work would be when I was still a student. As it turns out, my studies and passion for other primates brought me to a different kind of protection—protection for primates who are held captive, who are used and abused in labs, and who are harmed by the ugly global trade that captures and breeds primates for this terrible fate.

One of my colleagues at PCRM, Dr. Hope Ferdowsian, provides assessment and care for human trauma and torture survivors. She notes that psychiatric disorders often follow significant, repeated, and/or chronic traumatic experiences for humans. Chimpanzees (and other animals) who have been exploited by laboratories suffer such purposeful cruelties as premature maternal separation, adverse rearing conditions, social isolation, prolonged captivity, sensory deprivation, repeated physical harm, and intense pain and suffering. Hope and I are now working together to study similarities linking human beings and chimpanzees.

Despite striking similarities between human and nonhuman primates, we seldom use the same terms to describe or methods to study mental-health conditions across species. Instead, scientists rely on lists of abnormal or nonadaptive behaviors and traits in nonhuman primates known as ethograms. Items from these lists are measured according to their rate, duration, intensity, etc. Little attention is given to cause, development over time, self-expression, the way behaviors and traits group together into recognizable syndromes, or mechanisms such as the neurological and physiological underpinnings of the external phenomena we observe. Even less attention is paid to pondering the morality of exploiting other individuals in ways that lead them to abnormal and maladaptive behavior. Fortunately some scientists, such as Martin Brüne, Signe Preuschoft, and Gay Bradshaw, have called for a new way of thinking about the psychological lives of great apes and other animals.

PCRM also takes a novel approach to examining nonhumans, extending methods already used in human medicine. In these studies, knowledgeable informants (typically parents or caregivers) provide detailed responses to a series of questions about children or others who are unable to articulate their own experiences. This approach acknowledges the principles of guardianship and care that are central to sanctuaries and avoids any direct impact on the chimpanzees, who are especially vulnerable because of their presanctuary experiences in laboratories.

Using this method, we are studying traits and behaviors that indicate certain psychological conditions, such as depression or post-traumatic stress disorder (PTSD). Though they are no longer in cages, we are finding that many chimpanzees continue to experience depression and anxiety disorders, including compulsive behavior, as a result of previous exploitation in laboratories. If we find that these psychological conditions are common among chimpanzees previously housed at labs (or subjected to other forms of trauma and stress) but now in sanctuaries, we will have uncovered an epidemic of psychological disorders among captive chimpanzees. Understanding such psychological symptoms in captive primates is essential to implementing protection and providing appropriate care.

The pain and suffering of vivisection are directly linked to the ugly international trade in primates, which is devastating both individuals and species around the world. It may appear that primate suppliers in nations like China, Vietnam, Mauritius, and Cambodia bear most of the ethical burden for the primate trade, but both are profit-driven enterprises, and vivisection drives the trade in primates, not vice-versa. Indeed, the primate trade affects all of us, regardless of whether our respective nations buy, sell, farm, trap, transport, study, or test these hapless creatures.

At a primatology conference in 2008, a spokesperson from a Caribbean monkey farm referred to primates as a "renewable resource" or "asset." This cold rhetoric demonstrates that dealers view monkeys as mere commodities in a global marketplace—akin to recycled plastic or coal—rather than recognizing primates as a fragile part of threatened ecosystems or individuals who need and deserve our protection.

Those who defend primate vivisection cite strict government rules, regulations, and monitoring in the hope of assuaging a public that is increasingly critical of such experiments. Meanwhile, formal queries and undercover investigations consistently reveal violations of rules and regulations and a host of dubious practices. Primates who become the property of trappers or monkey farmers (who breed primates) are passed on to marketers, wholesale operators, and distributors, then eventually become unwilling subjects in experiments where they are poisoned, mutilated, and dissected. Some live in tiny cages for decades, surviving multiple procedures before they are killed. Some chimpanzees have been imprisoned and tortured for fifty years. Of these many exploited primates, few end up in sanctuaries because of lack of funding and space. Their futures are in our hands.

Speaking Out, Speaking Up

While investigating violations of the Animal Welfare Act in a government-funded experiment for PETA, I followed a case involving a group of rhesus macaques for nearly two years. As part of my investigation, I requested veterinary records for about a dozen monkeys (under laws that require open records, the state equivalent of the Freedom of Information Act). The files for two particular monkeys, Patrick and Brigit, were particularly thick, full of evidence of their hard lives.

Both were forced subjects of invasive brain experiments. Both had holes cut in their skulls that were fitted with metal guide tubes. The tubes held electrodes that were inserted into their brains during experiments to study brain activity. Patrick and Brigit also had bolts drilled into their skulls and a metal coil implanted in one eye. Four or five days each week, these individuals were strapped into a chair in full-body restraint with their heads bolted in place. Once immobile, they responded to lights on a screen, and the metal coils implanted in their eyes tracked their gaze.

For the three years prior to the reports, Patrick was able to reach through the bars and touch another of his kind for only a few months. One of the monkeys he lived near, named Shrek, was killed in an experiment. Another neighbor, Hercules, was sent away and blindfolded for a month, then reassigned to a different location. Except for these fleeting neighbors, Patrick lived isolated in his cage.

When Patrick "worked" in the restraint chair, he was rewarded with juice. To force him to work, vivisectors restricted food and water beforehand, assuring that he would be hungry and thirsty when taken from his cage. Even so, Patrick gained a reputation for not working the way his captors preferred, so his access to fluids and food was limited even more. Charts note that Patrick drank his own urine. Sometimes he reached between the bars and ate crumbs and other debris off the floor. Noting this behavior, experimenters moved him to a new location where he could not scrounge scraps.

In 2007 Patrick became so emaciated that university veterinarians had to give him special care: extra food, protein powder, and vitamins. Records show that he was thin as a rail; his spine and hips jutted out under his skin. Additionally, as a result of social isolation and the repeated trauma and suffering caused by captivity and experimentation, Patrick had developed severe psychological disorders. For example, his emaciated state was particularly conspicuous because he had picked out much of his own body hair. According to his official records, much hair was also missing from his tail.

Like Patrick, Brigit spent her life hungry and thirsty in the laboratory, but she suffered even longer. In 2007 she turned fifteen, having endured more than a decade of experiments, during which time she has lived alone in a cage, never grooming or huddling with another monkey for comfort or reassurance. In July 2002, when Brigit was just ten, a veterinarian noted that she was underweight and suffered from muscle atrophy—she had effectively no muscle mass along her spine or on her thighs. She was so malnourished and dehydrated that the veterinarian recommended more food and subcutaneous fluids. Despite her dire condition, records indicate that the vivisector who controlled Brigit's fate opposed the recommendation. Consequently, she received only seventy-five milliliters of fluid. Brigit gained a little bit of weight in the months following her tenth birthday but remained painfully thin.

Since 2002 Brigit has continued to suffer from dehydration, skin conditions, eye infections, and weight loss. Indeed, during the last six months of 2005, she lost 25 percent of her body weight. Gaps in Brigit's veterinary records leave unanswered questions. After a long pause, records in 2006 suddenly indicated that Brigit was blind in her right eye, but there is no explanation about when or how this occurred. Studying vision using a half-blind animal seems to be a futile enterprise and likely calls into question the validity of any data. In May of 2007, months after I began working on this case, the most recent records were released. Brigit's veterinary records noted that her left eye was swollen and oozing and the eye coil had therefore been removed—but there is no record of this surgery in the file.

While records confirm that Patrick was eventually killed, Brigit's fate remains unknown. It is, however, certain that what happened to these two primates was cruel and unnecessary.

Shifting Perspective through Advocacy

An important part of my work and advocacy on behalf of nonhuman primates is to try seeing things from their perspective and attempt to describe people, places, and things, as well as actions, causes, and consequences, from their vantage point. After many years, this has become second nature for me, so much so that I am struck by how often the point of view of nonhumans is not apparent to humans.

I was recently at a conference where a fellow presenter discussed animal art. Along with famous examples of great apes painting (and expressing their sense of aesthetics in myriad ways), the presenter described painting as enrichment at a primate vivisection laboratory. She showed pictures of monkeys in cages, clamoring to reach an enrichment device. She went on to extol the benefits of enrichment independent of context, ignoring the fact that these monkeys were being exploited for experimentation and would be killed prematurely.

How can vivisectors, universities, industry interests, and others with power make such obvious facts invisible? When does an audience listening to such a talk become part of the process of making animals, their experiences, and their suffering acceptable? In a world such as ours, how can we come to see and hear animals suffering in spite of such pervasive invisibility?

Maybe policy makers, researchers, regulators, and the public tolerate exploitative practices such as animal experimentation based on misconceptions, most notably the belief that chimpanzees, macaques, and other nonhumans are unharmed by research. Evidence about the pain and suffering—including the psychological impact—is well hidden from the public eye, yet this information is important if we are to improve public understanding and change lab practices.

These are critical questions for animal advocacy but also for examining any speaker praising enrichments in a lab. As it turned out, I had filed a complaint against this facility for violating the Animal Welfare Act (the lab was eventually cited and fined). Not only were employees at this facility cruelly exploiting nonhumans, but they were also evading the very limited laws protecting these animals. Nonetheless, I have no doubt that some monkeys did use the enrichments. What else is there to do? What was striking—at least to me—was what the speaker didn't describe: tiny cages filled with young monkeys with no adults in sight. Moreover, the speaker seemed unconcerned about what lab technicians and scientists would do to these primates later that month: how they would suffer, and how many would be killed by the end of the year.

Judging from the questions and comments at the end of the talk, the presenter's pictures and words led few people to these perplexing questions, perhaps because they do not know about life in laboratories. Still, it seems reasonable to expect people to ponder the lives and plight of these caged individuals. The most striking illustration of the conspicuous absence of animal perspectives and experiences—and what convinced me that both the presenter and the audience were actually unaware of the larger context of animal experimentation—was a single sentence near the end of her talk. The presenter explained that some of the paintings made by the monkeys in the cages were sold to the public to fund experimentation on those same monkeys. Not one person seemed shocked. I had recently watched a movie about Henry VIII and recalled that Anne Boleyn was required to pay the executioner to cut off her head.

Concluding Thoughts

The practices of primate vivisection—of all vivisection—and the arrogance of its proponents are deeply troubling, even morally repugnant. As a primatologist, I find primate vivisection intellectually indefensible. Everything that is done during every step of the process is inimical to our modern understanding of primates. We know that they are bright and sensitive animals with a long period of infant dependency and early development and complex social needs across their lifetime. We have evidence of psychological suffering caused by captivity and experimentation. Despite this knowledge, every aspect of primate vivisection is equally ugly:

- Primate vivisection starts in some remote forest when the trap door slams shut behind a few animals who are forever separated from family and friends.

- These individuals are tossed into the back of a truck and taken away from their vast forests to a distribution center or farm where they are imprisoned. Once sold to the United States, they are put into a crate and loaded into the belly of an airplane for a long intercontinental flight.

- After a terrifying journey, these wild monkeys are placed in sterile quarantine facilities for one month.

- After quarantine, they are moved to a new compound or trucked to another state, then placed in a three-foot-square cage (roughly the size of a kitchen cabinet), where they live alone—likely for the rest of their lives.

- The monkeys are then prodded, infected, poisoned, and otherwise exploited for experimentation.

- One day they are finally captured for the last time, then strapped down to be killed. Some of their organs or other body parts may be cut away while they are still alive and preserved before the rest of their bodies is thrown into an incinerator.

- Just a few years later (the minimum records-retention requirement under U.S. regulations), their existence can be erased when their personal record of torment and suffering is destroyed, including their veterinary file. Gone. Invisible. Forgotten.

Every single aspect of primate vivisection is not only inimical to what we know about these amazing animals but is also opposed to what we know about our own humanity. We diminish ourselves when we are cruel and ugly, when we do what is harmful, painful, and terrifying to those who are helpless in our power.

My personal experiences in an animal lab and the fundamental distinction that I found while living and working in Madagascar—seeing primates alive and free in their natural habitat—has shaped the way I understand these individuals and also the way I view primatology, my personal academic discipline. I no longer consider people who experiment on primates as primatologists, just as I do not consider myself an engineer simply because I use a computer or telephone. Nor was I a caregiver when I held that title and weighed primates in a lab or cooed, delivered food, or distributed toys. I formed bonds with monkeys like AB, but these were not healthy relationships; care was not my singular priority precisely because I was part of a system of institutionalized abuse, as are those who continue to handle primates in labs, trade, or entertainment.

Primatologists are quite different. They study primates as individuals who have value and worth in their own right. As with other professions, primatologists are ethically accountable to those who are the basis for their work and intellectual inquiry—nonhuman primates. "First, do no harm": I hope that someday all primatologists will maintain this ethic and those who do not will be shunned by their peers, or better yet, prosecuted and punished.

I feel fortunate to have parlayed my love of animals first into an education that centered on them, biological anthropology and animal behavior, and then into a career as an animal advocate. My personal journey to animal advocacy wasn't the shortest distance between two points; my path remains an ongoing adventure. I mention this for two reasons. First, the threats and challenges facing primates—from global warming to lax international laws—are dynamic. Thus, efforts to work on behalf of primates must be attentive, agile, creative, and responsive. Second, the need for animal advocates who want to work on behalf of primates (and many, many other animals) is urgent, and those who are passionate about animal advocacy should know that innumerable paths lead to this end. Imagine me—who studied and worked in laboratories where people imprisoned primates and experimented on them—eventually taking a job with an organization such as PCRM, working to protect animals and promote alternatives to animal research, exposing and terminating laboratory abuses.

Advocacy is demanding both intellectually and emotionally. Dr. Judith Herman, a world-renowned expert on trauma and recovery, explains why working with survivors of abuse and trauma is often difficult:

> To study psychological trauma means bearing witness to horrible events. When the events are natural disasters or "acts of God," those who bear witness sympathize readily with the victim. But when the traumatic events are of human design, those who bear witness are caught in the conflict between victim and perpetrator. It is morally impossible to remain neutral in this conflict. The bystander is forced to take sides.
>
> It is very tempting to take the side of the perpetrator. All the perpetrator asks is that the bystander do nothing. He appeals to the universal desire to see, hear, and speak no evil. The victim, on the contrary, asks the bystander to share the burden of pain. The victim demands action, engagement, and remembering (Herman 1992, 7–8).

Whether we are advocates for nonhumans only once in a while or regularly, whether we speak out for a monkey imprisoned in a laboratory or one who struggles for survival in a shrinking forest, we choose to share that burden of pain. And, at least for that moment, we choose to see, hear, and speak against evil and thus make visible exactly where we stand. Uncovering, interpreting, and telling the stories of exploited animals are part of my attempt to honor abused individuals and an intrinsic part of my efforts to make the world a safer, more just, and more compassionate place for all animals, including humans.

This essay is dedicated to the memory of AB and to hope for Brigit.

Part III

Sanctuaries

11

Born to Be Wild

Jungle Friends Primate Sanctuary

Barbara G. Cox

During infancy nonhuman primates are cute and lovable, but their endearing ways fade as the call of the wild grows more insistent. At just a few years of age, natural hormonal changes bring out aggressive behavior encoded in their DNA—behavior that can't be tamed or trained out of them. When they reach pubescence, monkeys rebel much as human adolescents do. Only their means of expression differ. Human teenagers shout and slam doors; monkeys scratch and bite. And, unlike most humans, they don't leave their wild ways behind when they grow older.

The lives of captive monkeys are tragic from the start. Monkeys imported to North America from the rain forests of Central and South America are stolen as infants from their mothers' backs. Hunters working for the exotic pet trade shoot the mothers from the trees where they live, then take the babies from their lifeless bodies. The infants are transported north in boxes and crates to be sold.

Almost all primate pets born in captivity come from breeders who warehouse fertile females, pair them with males long enough to produce offspring, and sell the infants in exotic pet markets. As soon as babies are old enough to survive on their own, they are ripped from their mothers' arms. Anyone who has witnessed this never forgets the cries of horror and grief.

The stories of captive monkeys usually have heartbreaking endings for both the monkeys and their owners. Eventually these individuals become disruptive. Many monkeys in the exotic pet trade are sold over and over again. Some are fortunate enough to find owners who place them in sanctuary homes. Parting with a beloved monkey leaves some people feeling sad and guilty, but in the long run, most realize that their pet deserves a second chance at life. Others are glad to see the tribulations of primate ownership come to an end.

How Jungle Friends Primate Sanctuary Began

Kari Bagnall is the founder and director of Jungle Friends Primate Sanctuary, located on twelve acres of wooded land in north-central Florida. Until her early forties, Kari was a successful interior designer in Las Vegas. Her life changed in the early 1990s, when her boyfriend bought a four-month-old capuchin monkey from an exotic pet dealer. The monkey, named Samantha, had a smooth, pink face surrounded by a ruff of soft dark fur. When she was being held or talked to, Samantha watched Kari with wise eyes.

By the time Samantha reached her first birthday, Kari's boyfriend could no longer tolerate the monkey's antics. Since the little primate's constant pranks had exhausted his patience, he proposed selling Samantha. Kari told him she couldn't do that. Samantha was like a baby, clinging to her around the clock. So the monkey stayed, and the boyfriend left.

Kari thought she had the perfect pet. She and Samantha went everywhere together. Unfortunately, their welcome soon wore thin at grocery stores, movie theaters, and shops. They were asked to leave. They were also escorted out of design showrooms, where Kari was working. Samantha was a menace in clients' homes. As soon as Kari put up a drapery, Samantha tore it down. Worse yet, the monkey bit clients who annoyed her.

Kari decided that Samantha needed the company of her own species. Putting one misguided foot in front of the other, she purchased Charlotte, a little capuchin sister for Samantha. This new capuchin didn't solve Kari's problems; she doubled them.

At first things went well. With her usual energy, Kari turned her home upside down for the girls. She spent a fortune on landscaping, laying a carpet of thick grass. She planted loquat trees so the monkeys could grab a fruit snack when they felt like it. She installed misting devices to keep the monkeys cool through the hot Las Vegas summers. A sculpted elephant fountain sprayed water into a shallow pool where they could wade and play. Kari even had holes cut in the walls and ceilings of her house to install runways from room to room. The tunnels and runways extended into the yard, and intercoms echoed throughout the house.

The monkeys even had a private room, complete with TV and radio. Being an interior designer, Kari made the house a showplace. Rain-forest wallpaper complemented the matching drapery and bedspread. Mosquito netting draped a four-poster bed. The lights were rigged so the monkeys could turn the switches on and off, and toys filled the room. No spoiled children ever had it better than Samantha and Charlotte. Kari was quite proud of herself.

However, her sense of motherly devotion was shaken when the busy monkeys began taking her perfect house apart. Samantha threw the TV across their room, and Charlotte tore down the wallpaper. As a team, they ripped up the draperies, ate holes in the walls and ceiling, dismantled the bed, and nearly hung themselves on the mosquito netting.

When things got boring indoors, they went outside. They stuffed pebbles in the elephant's trunk, blocking water flow from the fountain. They pulled up the grass and shredded their shade cloth. There was even occasional bloodshed at Kari's house. More than once, she had to drive a friend or family member to the emergency room to treat monkey bites.

Kari knew that the destruction had to stop. She decided to educate herself about raising monkeys in captivity. In the course of her research, she learned some brutal facts about the exotic pet trade. Her first reaction was anger and indignation. Then she decided to do something about it.

Kari made it her mission to educate the public—or, more precisely—children. Because she served as a court-appointed advocate for abused and neglected minors in Nevada at the time, she had a ready audience. Kari talked to the children about the disappearing primate habitat, the problems of human encroachment, and the need to preserve the world's fragile ecosystems. Samantha and Charlotte came along. The monkeys were as curious about the children as the children were about the monkeys. Still, the experiment backfired. The question the kids asked most was, "Where can I buy a monkey like yours?"

So Kari changed her strategy. Instead of bringing the monkeys to the children, she brought the children to the monkeys to demonstrate the problems of raising primates in captivity. That didn't work, either. The children still wanted their own pet monkeys. At this point, Kari realized that she had to practice what she preached.

Jungle Friends Opens Its Doors

In 1999 Jungle Friends welcomed its first residents—Samantha, Charlotte, and eleven other monkeys. Some of the monkeys who joined Kari on her trip east came from disenchanted owners. Others came from a breeder going out of business, who planned to auction his remaining monkeys off to the highest bidder. One monkey came from the entertainment industry. The troupe flew from Nevada to Florida in a small Cessna aircraft, then drove to its new home on twelve acres of property in the country. Kari chose an area just north of Gainesville, which enjoys a subtropical climate much of the year—certainly more hospitable to monkeys than the harsh Nevada desert.

Spacious, wire-mesh habitats awaited the newcomers. As the monkey population expanded over the following months and years, more mesh enclosures were added. They were built around existing trees and shrubs, but more plants were added. Ropes, hammocks, perches, and swings were installed to create a more natural environment. Underground lines were dug for water and electricity. Small prefabricated buildings were purchased and connected to the habitats so the monkeys could go indoors for warmth and privacy.

Jungle Friends could barely keep up with requests for permanent placement of unruly and unhappy pet monkeys. Over the next ten years, the sanctuary's monkey community grew to 120, including capuchins, spider and squirrel monkeys, marmosets, and tamarins.

Although each monkey came to Jungle Friends with a unique story, common themes emerged over time. The staff found that monkeys retired from the entertainment industry or exploited as pet-shop displays suffered from chronic anxiety due to prolonged exposure to the unwanted attentions of curious humans. Some were injured or maimed, bearing evidence of cigarette burns and amputated fingers. Often their teeth had been pulled to prevent them from biting their owners and handlers. Fortunately these resilient creatures did well in the Jungle Friends community. Their personalities made an about face in the sanctuary, an environment similar to their native rain forests. New residents made friends with other monkeys and before long were jumping from limb to limb, chattering with their neighbors.

Sadly, monkeys retired from research laboratories had the most trouble adjusting to sanctuary life. After being shut in cages for decades without a glimpse of sunlight and used as the subjects of stressful and often-painful experiments, they feared anything that moved. Their social life had been limited to interactions with scientists who were trained to take a detached attitude toward research animals. These unfortunate monkeys were wary of intimacy, even with their own species.

Monkeys who had been purchased as pets were also sad and difficult cases. Those whose teeth had been extracted could not eat properly and became malnourished, making them prone to metabolic disorders such as bone disease and diabetes. All suffered from psychological problems to one degree or another.

Stories told by former monkey owners are hauntingly similar. At first the owners dream that their monkeys will become lifelong companions, but their hopes gradually disintegrate, along with their home lives. In the end, the owners must give up these impossible pets.

Kari Bagnall loves welcoming former pet monkeys to Jungle Friends, knowing the happy life that awaits them. At first the monkeys mourn the loss of their human companions and don't understand why they were removed from their homes and familiar faces. However, they soon find new primate friends. Former owners may not recover as quickly, suffering persistent guilt and sadness.

The Daily Grind

Primate sanctuaries are demanding taskmasters. At Jungle Friends, all 120 monkeys are fed daily, each from a personal bowl. The main dietary staple—fresh fruits and vegetables—must be cut or chopped by hand into bite-sized pieces. Food processors can't be used because they turn the food to mush, and monkeys like to handle and inspect each morsel before eating it. Food service is a complicated business because species-specific meals must be prepared for capuchins, spider and squirrel monkeys, tamarinds, and marmosets. In addition, monkeys with health-related dietary needs receive special meals.

Maintaining habitats is also a big job. Every mesh enclosure must be cleaned regularly to remove food debris and feces. To prevent dexterous monkeys from escaping, the wire in overhead passages must be checked regularly, and loops connecting habitats must also be frequently tested and tightened. Monkeys are remarkable escape artists. Media reports of monkeys on the loose in the United States demonstrate that these events are traumatic for all concerned.

Like humans, monkeys sometimes get sick. A monkey with an infection requires extra time and attention and may need to be taken to a veterinarian. Those with chronic illnesses receive daily medication or medicinal herbs from the sanctuary's clinic dispensary.

Metabolic diseases common to captive monkeys can be life threatening. Because monkeys who have been purchased as pets are generally fed the same diet as their owners, including junk food, these animals are prime targets for diabetes. A number of diabetic monkeys live at Jungle Friends and require daily insulin injections or oral medications.

Social interactions among the monkeys must also be monitored in case fights break out. Monkey quarrels can escalate quickly, sometimes resulting in bloodshed. Staff members rush to help when they hear cries of excitement or fear from one of the habitats and quickly separate the squabbling monkeys. All staff members carry walkie-talkies so that they can be summoned immediately when emergencies arise.

At night live-in staff members remain alert for monkey warnings of intruders. While a chain-link fence surrounds the property, break-ins are not impossible. Fortunately monkeys make excellent security guards, screaming when anything is amiss.

Cost of a Primate Sanctuary

Reputable primate sanctuaries promise permanent homes to the monkeys they accept—a commitment that can translate to forty or more years of care. Unlike dogs or cats, monkeys can't be put up for adoption to good homes. And because these monkeys haven't learned the skills necessary to live in the wild, they can't be sent back to the rain forests of Central and South America. The demand for sanctuary placement increases every year. Jungle Friends never has enough space for monkeys on the waiting list. Running an ethical primate sanctuary is costly. Because such facilities exist for the welfare of their residents and not as public entertainment, expenditures often outweigh revenue. Sanctuaries depend on generous donors to stay in operation. Some give money; others contribute materials and equipment. At Jungle Friends, the help of volunteer workers is also critical. Without them the sanctuary could not continue to operate.

In today's stumbling economy, contributions to Jungle Friends and other animal-welfare organizations have plummeted. As a result, the search for monkey sponsors and donors is more pressing than ever. To help raise money, the Jungle Friends website (http://www.junglefriends.org) promotes monkey sponsorship and sells gift items such as monkey paintings, calendars, and greeting cards.

Promise for the Future

On the bright side, primate advocates have been instrumental in forcing government regulators to crack down on the illegal pet trade. This should help reduce the number of primates requiring sanctuary in the future. The overall demand may be balanced, however, by an increase in requests for placement from research facilities, which are beginning to retire primates, rather than kill them, when their usefulness as laboratory subjects ends. Some universities report that they are getting out of the monkey business altogether because of public pressure.

One of the most ambitious missions of Jungle Friends Primate Sanctuary is to educate the public about the needs of primates and the perils of keeping monkeys as pets. Kari Bagnall's goal is to convince people that monkeys don't belong in cages wearing cute clothes. They belong in the rain forest with others of their species. Monkeys are born to be wild.

12

Loving and Learning

Deborah D. Misotti

I find the heart and soul of every living creature in his or her eyes. There, I've said it. It has taken me many months of angst to bring myself to assigning this decisive statement to paper.

I have had many lessons, and many teachers, in my lifetime. I find myself marveling—when I think about it—that very few of my best lessons have come from humans. In fact, in retrospect it seems that the most valuable instructions have always come when I least expected them and in the strangest of circumstances.

The experiences of my life have culminated in a gift of intuitive knowledge accumulated and valued through my unique education, which cannot be replicated within the hallowed halls of academia. The ivory tower of education has resulted in some of the things I value in my life, but the experiences I have had elsewhere are the gifts that I treasure most. This life I now live is at the end of a journey I have traveled for more than half a century. Now—in the luxury of hindsight—I can reflect on lessons from the past.

Corky, July 1974

I sat on my porch in this wonderful old farmhouse, looking out over fields of corn to the pastures beyond with cattle, horses, and a sky so blue it seemed like an ocean above.

I came home from the hospital today with empty arms and a stunned heart. My infant son did not come with me...he could not. He died yesterday, a result of an oral teratoma, such a rare and frightening birth defect that there had only been one other recorded case in history, eight years earlier, in Cairo, Egypt. No one had known what to do...they still did not.

So my Jason died, and I was left with empty arms, a broken heart, and a very uncertain future. Tom, my husband of five years, had beggared himself to give me an anniversary present we could ill afford to keep. We had no medical insurance, and it would take ten years to pay for our son's birth and inevitable death. Nonetheless, Tom gave me an Irish setter puppy, which I named Garland's Lord Corcoran—Corky. I had wanted this puppy since I saw the Disney movie Big Red when I was a small child. I watched him scamper—nose to ground—with my four-year-old daughter, Dulcy, running behind. She was determined to convince Corky to lift his head so he wouldn't flip himself endwise when his nose hit a root.

I felt numb.

Dulcy and Corky played together, forming the sort of bond that lonely young children create: an only child and a puppy who had been taken from a litter of twelve. Perhaps this wriggling red bundle of energy would become her brother in his own loving, silly way—her partner in childhood crimes.

I rose from my rocker and headed for my meadow—not truly mine but a beautiful spot my neighbor has kindly allowed me to call my own. It is a lovely, hidden place in his cattle pasture, nestled in rolling hills. I lumbered down the crest of the hilly pasture, pulled by my memory of a natural spring filled with water—icy cold and strong with the taste of iron—and tall trees. I remembered a fence line of multifloral roses sharing their sweet scent, wafting over smells of cattle, horses, manure, and the pungent odor of Maryland red-clay soil. I expected to find dragonflies and the perfumes of nature in this place where, perhaps, my heart might learn to feel again.

I passed cattle grazing over acres of rolling green. I shadow-walked the crest of the hill that led along the winding pathway to my sanctuary. Nature would help me; I knew this as I paused at the top of the steeply sloping path to my meadow, to peace.

I heard the lowing of a cow separated from her calf. My farmer friend, Gilbert, separates the cows from their calves around this time of year to sell them at the market to become hamburgers. Another cow joined the sad, lowing call, and I suddenly shared their great pain of loss. I dropped to my knees. Hot tears tumbled, and great racking sobs wailed my grief united with those other poor mothers who had lost their babies, their children.

I don't know how long I sobbed and screamed my anguish. At some point, Corky scampered up and licked my tear-streaked and swollen face, then lay down on my lap, allowing me to cuddle him in my raw, empty arms against my painfully swollen breasts. The sun was setting, and my new companions in circumstance—bovine mothers—were still sending their painful message of loss to the sky as I turned from the precipice and returned home, looking down into the eyes of my newfound son, Corky. (Lesson: Corky 101.)

Ruby, March 1997

It was a bright spring morning, and we had been invited to visit a particular nonhuman primate rescue project located within a commercial animal entertainment facility. This business was unique because it sponsored and financially supported a fledgling nonprofit sanctuary, in spite of the fact that entertainment was the primary goal.

Generally, entertainment facilities exploit young primates to bring in money from eager viewers, but through its nonprofit rescue project, this place provided for babies born there even after they grew up and were retired from the exhibits. Eventually the rescue project became self-supporting and moved into separate quarters, but until then the entertainment facility provided caging, housing for the project office, food, paid caregivers, and veterinary care and even used the captive primates in fund-raising ventures to create the sanctuary for these displaced individuals. I tell you this to make the public aware that some commercial animal entertainment operations and pet stores have a bit more conscience and take at least some responsibility for the sufferings they cause. However, they still exploit primates who are endangered and harmed through their actions, so I refuse to advertise for them and will not divulge the name of this facility. Nonhuman primates belong in the wild and can only be protected if they are not treated like commodities.

The world of animal entertainment is filled with young primates who entrance and delight the public—whether they wish to or not. Unfortunately, as these individuals grow, people do not find them as endearing. In truth they can also become dangerous, but they only have one place to go—breeding facilities, where human contact is denied and they and their reproductive abilities are exploited for science or the pet market. Scientists in white coats use these unfortunate primates again and again until a particular study finally takes their lives. The pain and misery they endure in laboratories, where they experience depression, confusion, sadness, and boredom, often creates irreversible psychological problems. They are also forced to produce babies for pet buyers, who constantly rob their offspring from them.

This Saturday Tom had worked during the morning, and I had anxiously awaited our scheduled afternoon visit to the primate rescue project. I was excited, and Tom was resigned. It would be a new experience for both of us, I reasoned, but he came along with less enthusiasm than I liked to see. Somehow this venture did not particularly appeal to him, though I could not understand his reticence. Nonetheless, he indulged me (as he often does).

At the project, we were introduced to several apes, young chimpanzees, and orangutans. To my delight, we were allowed to interact with them among the huge trees and lush vegetation behind the public facility. Their musky smells and ripe, spicy body odors assailed my senses. One young chimp—a female—bumped into me while swinging energetically from a tree limb. I was thrown off balance and cried out. She quickly reversed direction and looked deeply into my eyes. I was startled by her intelligence, cunning, and raw power. I stepped back—amazed—and she dismissed me, swinging back to her own world.

In that moment, I felt a brief stab of sheer terror that I had experienced only once before, as a child, when I encountered a copperhead poised to strike along a mountain trail. My father threw a rock, and the snake left as swiftly as the chimp did now. As a child, I had dissolved into tears of fear and relief. Now I was more seasoned, and amazement and exhilaration quickly replaced my fear. How astounding to witness such a wondrous creature in action, up close and personal!

Not long after, a young orangutan walked toward my husband with a deliberate stride and a large toothy grin (full of impressive teeth). The rangy red ape stood more than three feet tall and was rather imposing. Tom froze as the orangutan lumbered toward him, then reached up and strongly pulled on Tom's right shoulder, forcing him to stoop. The project director called out, letting us know that the ape simply wanted a piggyback ride. Tom, whose complexion was, by now, sheet white, obliged with only momentary apprehension. We still cherish the picture I snapped of that awkward, but wondrous, moment.

We walked back into the park with young apes following behind, Tom still carrying the orangutan piggyback, and entered a wonderfully lush jungle environment with large cages of monkeys and apes. The general public was milling about on the other side of the fence, but we were in the coveted area with the orangutans, chimpanzees, and other primates.

Then—out of nowhere—a woman appeared, holding the hand of a beautiful little rusty-colored orangutan, who smiled at me and reached out her hand. I linked hands with her and looked down into the most beautiful golden eyes I had ever seen. I felt as if I were falling deep, deep down into a world beyond. In her eyes, I found amusement, intelligence, and an invitation so palpable that I thought she had spoken: "Come, learn, become as one with me."

Ruby became one of my finest teachers and dearest loves. She was the kindest, most extraordinary friend I have known. My world was never the same again.

Ruby, Tom, and I shared only a little more than a year before we lost her to an enemy she fought most valiantly, a disease many individuals have fought—cancer. Her life—short though it was (only five years)—was rich with friendships and love. All of the volunteers who knew her at the project learned from her; she was a demanding teacher. She was a termagant—a holy terror—but also a kind soul who gifted us with hugs when she could see that we were in need, bites when we disobeyed her rules, and many, many hours of laughter and happiness. With Ruby I experienced all of the emotions possessed by the human spirit, and I reveled in every shared moment. The knowledge that she shared flows through me to the people I speak with today. In this way, she is always with me. (Lesson: Ruby 101.)

Abby, October 1997

I had been volunteering with the primate project for four months. I very much wanted to work with gibbons, but I was told that only a few people were allowed to try to connect with the gibbons, and very few succeeded.

I was persistent, and the curators finally acquiesced. For several weeks, I courted two gorgeous creatures—one gold male and one black female, both white-handed or lar gibbons. All of my interactions came from the outside of their cage, however; it is dangerous to enter what gibbons consider to be their space. Gibbons are territorial and will defend their homes. They accept few humans into their territory.

I found that even standing outside a gibbon's cage can be dangerous. I was rebuffed repeatedly, even physically injured. Their stringent rejection of my presence was evident: pulling my hair, tearing my clothing, scratching, pulling me up hard against the bars of their cage and making loud popping smacks, which left large bruises. Nonetheless, I came back, determined to gain their trust. As in the past, the black female—Abby—grabbed my bangs and rapped my head sharply against the bars of the cage, painfully bruising my face and forehead. I walked away, rubbing my head, and looked up to see an old man watching with his arms crossed over his chest. I immediately stopped rubbing my head and put on a sheepish grin. He kindly asked, "Did she hurt you?" I told him that the hurt was not here—in my head—but rather here—in my heart. I felt so close to them, yet they would not allow me to be their friend. He shook his head and asked me why I loved the gibbons. My answer was simple: "Because they sing."

It was true. The musical voices of the gibbons lifted my heart to the heavens more than any choir. Gibbon song has been a siren's call for me since I was a small child wandering the zoos of Baltimore and Washington, D.C. My parents always knew where to find me if I wandered off—I would be listening to the gibbons.

The old man smiled and nodded, but my mind shifted, and I was suddenly struck with inspiration. I turned without another word, ran and jumped onto the rock wall around the gibbon cage, and began softly crooning a lullaby.

The song I chose was one I never expected to sing again—the lullaby I had sung to my son the day he died. I cried as I sang, and Abby responded. She raced across the cage, and I braced for her violent blow, which I expected would send me crashing to the ground. But she did not strike me. Instead, she reached out her hand and wiped a tear from my cheek, then tasted the salty drop. She repeated her taste test, then started to croon softly along with me, wiping my cheek, sharing my pain, and becoming my dearest, closest, and most loving sister. I vowed in that moment that I would rescue her from the eyes of the public, which I knew she hated. (Her mate, Sandy, loved clowning for crowds, but Abby was private and did not like humans ogling.)

Every day thereafter, Abby and I sang together. She started singing when she heard my Jeep heading down the two-mile road to the primate facility, and I sang along at the top of my lungs, in spite of gawking visitors. Our bond grew stronger song by song. Not everyone is privileged to sing with a gibbon, and to me this was a gift beyond measure.

One of my deepest regrets is that I was not able to save Abby from the entertainment facility where she lived. Tom and I worked frantically to set up a sanctuary so I could bring her home before she was sold, but it was not to be. She was purchased by a breeder, who ultimately left her caged—outdoors—during Hurricane Wilma. She died of a heart attack when a large tree fell and crushed her cage with her inside. I failed her, and I lost her forever. The dream of one day bringing her to our sanctuary, so lovingly created with her in mind, vanished. I live with this pain every day, and each memory still prompts my tears. (Lesson: Abby 101.)

The Sanctuary, September 2001

Tom and I, now married for forty years, decided—after much volunteer work at exotic animal rescue projects and animal entertainment facilities—to create our own sanctuary, a rescue project. The definition of sanctuary is "a place of refuge or asylum." Besides offering safe haven, our sanctuary would incorporate ecoeducation, encourage a global vision, and heighten awareness of the possibilities nature provides to ordinary people. We had learned much over the years from living with the Earth and its creatures. Through our sanctuary, we would seek to alter the consciousness of the people of south-central Florida simply by living close to the Earth with rescued nonhumans. Perhaps we could encourage vision in young people, even stimulate lost vision in adults. As children we all have dreams. We would become an example of the way—even after half a century of life—people might still live the dream they envisioned in their youth.

As we planned for our sanctuary, I spent hours on the phone or computer learning more about nonhuman primates. Some people "talk sports," but I was always "talkin' monkeys," as Tom put it. We decided to call our rescue the Talkin' Monkeys Project, and so our sanctuary was born (Lesson: The Talkin' Monkeys Project, Inc. 101).

Chi Chi, July 11, 2008

I watched with trepidation as the van pulled into our compound, delivering another small soul who would rely on us to make her world stable and safe. I anticipated her fear—the confusion and resignation this poor little gibbon was experiencing. Chi Chi had been transported many times during her life of eighteen years: new people, new cages, new neighbors, new experiences...again and again. It was my hope that our sanctuary would be this beautiful black white-handed or lar gibbon's final earthly destination.

Chi Chi deserved to be housed in a large cage with chutes and towers and an attached, secure building so that she would have heat in the winter, cooling breezes during the summer, and a protective roof. We provided open, sunny shelves where she might sit and relax, poles on which to brachiate, fire hoses where she might climb, swing, and sleep, and plenty of food, fresh water, and companionship when and if she desired them, as well as areas to escape for solitude. Perhaps even some bamboo to eat or preen or play with. But most of all, Chi Chi deserved to be treated so that she could be a gibbon without the constraints of human greed, ownership, and intrusion.

Chi Chi came to us from a breeding facility where she was constantly producing babies, yet never allowed to be a mother. Her babies were sold to private owners, other breeders, laboratories, and zoos. A gibbon baby usually sells for ten to twenty thousand dollars. At Talkin' Monkeys, Chi Chi would not be forcibly bred. There would be no influx of breeding males or constant human observation to see if she might mate with some stranger. Breeders care about gibbons for only one reason—their babies. Gibbons are the only monogamous nonhuman primates: both parents nurture and raise their infant—if given the opportunity to do so. Chi Chi had never been given a companion to live with, and she had never been allowed to raise and tend her young.

On the day of her arrival, she started a new life of leisure, a life of unconditional love and care, a life with freedom of choice where food was not served on a schedule mandated by the clock, and she could choose to play with toys in her surroundings or sit in the sun. At the Talkin' Monkeys Project, clocks are foreign. Life is not governed by the ticking of a mechanical faceplate. Food is available throughout the day; primates receive no less than five feedings, more when time allows. We would be around if Chi Chi needed us, but we would not intrude in her new life. She would eat and play as she pleased. We hoped she would learn to be at ease with us over time. I trusted that lessons would be learned from both sides of the cage, and I looked forward to this new education. (Lesson: Chi Chi 101.)

Epilogue

Dulcy has grown into a lovely young woman with a life and family of her own, and her canine brother, Corky, passed on when Dulcy reached her teens.

I have had a rich life, enhanced by the myriad nonhumans I have been fortunate to know. Yet I have always longed to recreate my meadow—or at least that feeling I had there—which offered such peace in moments of stress and loss.

Over the past few years in South Florida, nature has changed the topography of our land. Hurricanes, tropical storms, drought, and our sanctuary have created an unusual and beautiful area complete with monkeys, horses, dogs, cats, goats, and volunteers who share our lives.

As a sanctuary, we continue to rescue anyone who needs us—human and nonhuman—even if only for a little while. Needy individuals come to us wounded—sometimes physically and sometimes in spirit—often lonely and wary of strangers. Sometimes the domestic animals—dogs, cats, horses, goats, and so on—leave for new homes, healed and happy with a bright life ahead, filled with new people and new experiences. Sometimes our friends leave us unwillingly, and we grieve for the loss of souls we have loved. These are the hardest partings, but we know, in our hearts, that we have loved each soul while it was here, either as a volunteer or a friend who became a member of our exclusive little family. Only the primates that come to Talkin' Monkeys have found a home for life.

I have learned much from many teachers beyond those I have described, but the treasure beyond measure has been the gift of friendship. I have learned to cherish these gifts and share them in lectures, conversations, and classrooms. I live close to the Earth and its creatures in peace and harmony. I hope this will be enough to leave behind when I am gone.

Each morning I smile as I awake to a wondrous chorus of gibbon heart song, but as soul satisfying as this is, I am not fully awake or alive until I look into a pair of those round, golden brown eyes set in a little black elfin face ringed by a thin circle of white hair...the eyes of a gibbon. My wish is that others will look deeply into the eyes of those they meet: look for their souls and seek the lessons they offer. This sanctuary full of monkeys, gibbons, and bamboo is my life's dream. What is yours? It is never too late for dreams to come true.

Afterward

As this anthology was headed for the publisher, our sanctuary was approached by the owners of a young pet gibbon named Webster, whose vocalizations were causing a problem in their residential neighborhood. The owners planned to sell him to a zoo in Connecticut or an entertainment facility in Miami until they heard about our sanctuary. Ultimately, they decided to forego the monetary benefits of selling their gibbon and placed him with us—where he would be the happiest. Our joy was compounded when we discovered that Webster is the sole surviving offspring of Abby, the gibbon for whom our sanctuary was created. What we could never do for Abby has become a reality for her son—life in a sanctuary. Here Webster will be a gibbon, not an object of entertainment. The circle is complete; the hole in my heart, where Abby has lived for many years, is beginning to heal, and Webster is as safe and free as a gibbon can be, removed from his or her home (Lesson: Webster 101.)

13

Some Baboons in My CARE

Saba, Einstein, George Bush, Nathan, Snare-Boy, Tripsy, and Giovanni

Rita Miljo

The babies had settled into their sleeping quarters with bottles and blankets, and I had snuck out to have a few moments by the river after an unpleasant day. It was playtime for the Longtits and would be dark in ten or fifteen minutes. The troop was going full blast: youngsters were wrestling, cartwheeling, fighting and teasing the older girls. But the older girls didn't mind. With full bellies—after a good day—even Gretchen invited the odd youngster to take a ride on her back. Scruffy came to look at me intently, then very carefully touched my arm and started to groom me. Little David bounced around, then came storming up to displace Scruffy. Tripsy slunk by, looked at me, and gave me a quick lip-smack. Little Fred came charging along—now quite a big Fred—and put his arms around me.

I am in the process of discouraging such intimacies; he has become a huge male, who should not take such liberties. Thinking whether I should tell him to go, my heart nearly stops. About three meters away, a young elephant bull lumbers past, close enough to smack me with his trunk. I sit dead still and am grateful for little Fred's inclusive arms around my neck. It seems that the elephant associates me with the baboons, who do not seem to worry about him; they hardly move, and he walks on.

In South Africa, baboons are not very highly esteemed. In fact, they are despised, ill reputed, and persecuted. Somehow people seem to hate them for their intelligence, inventiveness, and gift for outwitting "superior" humans.

Traditionally, before European settlers arrived in South Africa, locals respected baboons, acknowledging their special cleverness. There are many South African legends and folk tales about baboons, and respectable local communities call them "our people" while other locals associate baboons with witchcraft and fear them. Unfortunately, European settlers brought guns, irreverent killing, and large-scale destruction of the environment, all of which persist in spite of a trend toward reaffirming traditional values.

The apartheid government wrote laws declaring certain species of South African wildlife to be "vermin," including baboons, vervet monkeys, bush pigs, jackals, and caracals. They reasoned that God was drunk when he made these creatures since they showed no respect for farmers and their livestock. They decided that destroying these species would rectify the problem. Strangely enough these laws still exist.

At CARE (Centre for Animal Rehabilitation and Education), we tend three types of baboons: those who have been orphaned, abused, or injured in the wild. Orphans are almost always the result of humans killing their mothers. Orphaned babies need special care. We raise them with all the love they need, then put them back into the wild. Horribly abused baboons are also sent to us, but all we can do is try to ease their pain by providing respite. We also have a group of wild baboons who made themselves at home on our little farm in 1992—the beloved Longtit family who teach my growing orphans what baboon life is all about.

Saba was one of our abused residents. She arrived some twenty years ago, a frightened young girl who had been welded into a forty-five-gallon drum. Her keepers left a small hole at the top so they could throw food in and a hole at the bottom so a witchdoctor could harvest her feces for muti (witchcraft medicine). In her misery and despair—living in this small drum—she had mutilated her arms with her teeth. I cannot even think what horrors she experienced.

When she arrived, we felt that she was slightly too old and frail to survive the stress of joining a troop. So she took up residence with Bafana, a young male with an injured foot, that had been caught in a gin trap. At first all went well, but then Bafana started bullying Saba, so we moved her into her own apartment next to other severely damaged baboons—ones who had been exploited in labs. She made friends with these baboons and became especially close to old Nathan, whom she groomed for hours.

One day we noticed that Saba seemed unable to swallow her saliva. We moved her to sick bay since I thought she might have a bad tooth or some other minor, old-age problem. Saba was delighted. Whenever she saw me, she talked to me. I was concerned because she was terribly thin, so I spoilt her with soft food, hoping to strengthen her so that we could tranquillize her and check her mouth. When we finally looked in her mouth, a third of her tongue had simply disintegrated—there was a fully fledged ulcer. We immediately took her to the vet, but there was nothing that could be done. Saba never woke up, and I would be so much happier if I were sure that there is a baboon heaven.

When I pass Saba's empty cage, I hear her enthusiastic greeting. She never forgot me, the person who freed her from such a merciless prison. She deserved heaven. Humans made most of her life miserable. How many millions of creatures suffer at the hands of humans? Did Saba find happiness at last?

Einstein was an orphan. She arrived at CARE as a strange-looking little girl with her hair standing up in all directions, or rather standing on end as if an electric current flowed through her body. Einstein's social rank was rather low, but this did not bother her. She was very clever; she certainly took full advantage of my soft spot for her. She was a very special little girl: enormously intelligent, soft, and loving.

Einstein's troop was eventually turned out into the wild, my last single-handed release. I lived with them for six months, watching to make sure the baboons would adapt to their new home. At night Einstein snuck into my tent, asked for her special milk bottle, and cuddled with me—it was winter outside. In the morning, she pretended she hardly knew me, reverting to full baboon behavior. After the six months were over, I knew she would be fine. She was the first baboon in her troop to give birth, and I could see she was happy.

Looking back, I suppose it was rather foolish of me to expect to climb mountains and race after baboons as I had done years earlier to assure that their readjustment to the wild was safe and complete. After all, I was more than seventy years old when Einstein's troop was released.

Nonetheless, it was a successful release, and eventually I went home, leaving the baboons to their life in the wild. Every six months we monitored their progress, and Einstein was usually the first one to make contact. She knew the sound of my car engine—my voice—and was there to greet me with her big smile.

I often thought of her and wondered how she was coping. One day I received a phone call from a young man who lived in the release area. (Public relations are so very important!) This young man, a farmer, remembered the story of a baboon release in his vicinity, and he went to the trouble of finding me to tell me about an odd baboon: a rather old female who had taken up residence in his tractor garage, slept on the tractor, and seemed friendly. I was curious. He called during a particularly harsh winter—the nights were below freezing—and something made me think of Einstein. It had been four years since her release...but why would she seek humans now?

I asked the young man to watch her and said that I was headed his way. It was more than eight hundred kilometers to his farm, but I knew I had done the right thing when I arrived. There was Einstein, comfortably installed on the man's tractor—out of the cold wind—grooming herself. I called her name and was rewarded with a sunny smile. She immediately came over to me, lip-smacking with pleasure, and started to groom me as if asking, "What took you so long?"

Einstein climbed into my car, and nothing on earth could have dislodged her. As I drove away, I noticed two young baboons watching intently, but when I approached, they fled. Were they Einstein's children? She was not the least concerned. And so we headed for home with Madam Einstein comfortably installed on the seat next to me.

Back at CARE, Einstein entered one of our bachelor apartments with a nice bed and blanket and contentedly munched bananas with a big smile. After four years in the wild with her troop, she seemed to say, "I am home; freedom is for the birds." Perhaps Einstein was concerned about the cold and her old age, or maybe she missed the comforts and goodies...or maybe she remembered an old friend.

It is hard to believe that anyone would shed a tear over the death of George Bush, but George was a very special—and much abused—baboon. She came to us as a middle-aged (or perhaps older) lady. From the beginning, she had precious-few redeeming attributes. I was told that she had been found in a small, dungeonlike enclosure where old cars had been dumped. We gave her a spacious enclosure next to a group of baboons and hoped that we could eventually merge her into this small troop. We put another old lady in with her, one who was in need of love and care. But George made one blunder after the other, some of which backfired.

It wasn't her fault. George had never had an opportunity to develop social skills. Most baboons overcome this through intelligence, but George Bush had no intelligence whatsoever, or so I thought. We removed George's companion, and she was on her own and happy with herself. The highlights of her life were tidbits of food, and she spent hours lying on the floor, tickling her behind with a stick. We wondered, was George Bush dreaming or scheming?

One day when I visited George, she was in her usual pose on the floor, but something did not seem right. I went into the enclosure and found her to be almost stiff: she had tetanus. We took her inside and administered medicine, but it was too late; George died.

George had a sad, joyless life, ruined by humans. I felt guilty that I had not tried to make her life a little bit happier. Instead, I gave her the name of the most stupid man in the world. It could have been different if she had been a happier and more charming baboon. But people prevented her from being that way. I should have stepped back and acknowledged the bigger picture, the misery that her life had been, the human miseries that had created the baboon I foolishly called George Bush.

Nathan had also been abused. He had been in baboon Auschwitz—he was used for medical research. He was the oldest lab baboon that we had adopted. From the beginning, he was always my special boy. He had such distinct features: a pronounced chin, dignity, and a measure of calm. Vivisection labs only work with fully grown baboons, and he had been at the lab for thirteen years; he must have been about thirty-one years old.

Nathan occupied the middle section of Nut Village, where we put baboons destroyed by humans. We placed Nathan with Rhona, Sybil, and Gwinnie: three old ladies. But it was Saba, on the other side of the wire, that Nathan loved. He spent hours leaning against the fence while Saba groomed him. Even at night, I noticed that they slept on the ground so they could huddle together, though a fence separated the two lovers.

Nathan was a bright boy who knew that his enclosure was safe from predators. The best feature of his home was a high viewing platform with a thatched roof, from which he could look far into the wildlands across the Olifants River. He loved spending time there.

One day I noticed that he had difficulty climbing the long ladder to his loft. Being a very old lady myself, I decided to try that ladder to see how difficult it was to climb and to check out the view. When nobody was around to watch me, I started up that ladder and was not yet halfway when I decided that it was high time to retreat to solid ground; my days of tree climbing were definitely over. My respect for Nathan grew—the old boy just did not give up!

I was convinced that Nathan could not possibly have any teeth left; we cut food into very small pieces for our oldies to be sure that they have as much soft food as possible. The resident vervet troop was delighted. Since they were small, they could always find some small crack to slip through into Nathan's kingdom and raid his delicacies.

One day a devoted CARE worker, Bennett, rushed in to tell me that Nathan was foaming from the mouth and could barely breathe. I was reluctant to dart him because of his breathing problem and immediately phoned the vet. Lizanne, our devoted baboon vet, arrived within an hour. We darted Nathan and found that he only had four teeth left and, much worse, he had severe pneumonia.

We made him comfortable in the sick bay, though I hated to put Nathan into a cage just like those contemptible lab prisons where he had wasted thirteen miserable years. Nathan must have felt betrayed. After he woke up, I tried to hand him small pieces of banana, but he severely scratched my arm and turned his back on me. In spite of medication, he did not improve, and after two days of nerve-racking moaning, I decided to take him back to the vet. Lizanne also feared that there was more than just pneumonia to worry about.

It is a terrible responsibility to make life-and-death decisions about other creatures. But when I saw Nathan's suffering and thought of what I would want if I were in his position, I knew the answer. I could not in good conscience—subject him to so much suffering and confinement in a hated, terrifying cage. He had become an old man, and his special Saba was gone. We decided to let him go to join her.

We darted Nathan, then Lizanne started to shave his breast for the final injection. As she worked, I saw her pause. She looked up, shocked. She had uncovered a number tattooed into the flesh of Nathan's chest: 756. He really had spent most of his life in baboon Auschwitz!

Snare-Boy had been injured in the wild. He was just one of many happy baboon youngsters produced by a good rainy summer after a long spell of drought.

I was not even aware of his misfortune until my coworker, Bennett, pointed him out. He was limping and had a bird-snare wire around his left lower arm. His limp worsened, yet he outmaneuvered all capture efforts for nearly three weeks; by that time, he could not use his arm at all, and he was miserable. Normally he would almost have sat on my lap, knowing I was safe. But he was in so much pain, and nobody could explain to him that we only wanted to help. We tranquillized him just in time. The snare had cut deeply into the flesh of his arm, but there was hope with the help of our wonderful vet. The swelling subsided, and he slowly started using his fingers.

From his sick bay cage, he could watch his troop and talk to them; it was heartbreaking to hear him call for them as they left. What a happy morning it was for him when I casually left his cage door open. He looked at me as if to say, "You are very stupid this morning," then shot out like lighting, straight into the arms of his friends. There was a big welcome, and I watched him cartwheel down the hill toward the river, laughing and chattering with his companions. Lucky little chap! Without help he would have died. And our reward was his superior smile, saying, "You see, I outwitted you eventually!"

I often check the Longtit troop to see if they have gotten themselves into trouble, or if humans have caused them any misery. On one such check, I found Tripsy, whose left arm was broken in two places. She was dragging herself along, probably bent over with pain from internal injuries. In spite of these overwhelming problems, she fled in panic when I approached. What had humans done to her?

I searched for this wild, injured baboon for days and eventually found her hiding in thick bush. She could hardly move and so had to accept my approach. I was amazed that she was still alive and began to feed her baby milk, bananas, and eggs. Her eyes told me how desperate and terrified she was. After a few days, she could drag herself away; I followed and gave her food. Her will to live was unbelievable. After some time, she started to move around again, but I had made it a habit to look for her in the morning.

Tripsy never trusted me or gave me the slightest encouragement or recognition, but she gratefully accepted my food for years. We had a standing agreement. We met in the morning out of sight of the troop behind the mamba kitchen. Tripsy enjoyed her food, especially her two bananas, until the troop discovered the food source, charged in, and drove Tripsy away. Most of the time we outsmarted the others, and Tripsy put on weight. She stuck strictly to our morning meeting times, and I knew that her well-being depended on me, that she would be waiting. But there was no intimacy between us. Tripsy had decided that humans could not be trusted, and she tolerated my interference only because she was dependent. I was sad that nobody in the troop took much notice of her, or tried to help or console her. On the contrary, I had to protect her and make sure that she could eat the food I offered.

During my younger years, I was fortunate enough to travel South African regions with a German researcher trying to find what he called "wild Bushmen" in Botswana. He was as passionate about Bushmen as I would later be about baboons. He had learned their language and told the most amazing stories about Bushmen, their way of life, the way they saw the world, and the way they lived and battled to survive in hostile surroundings. They live and move in family groups, and he noted that family members who were too old to move with the group were left comfortably settled with an ostrich egg full of water and some food. The well-being of the clan had to come first. I found this hard to understand at the time. Watching Tripsy, I came to understand the way the laws of nature worked and were applied by communities. But I did not need to leave her to her fate.

And so Tripsy lived for another five to six years, sometimes looking better than ever and sometimes not. She was a loner. I watched the Longtits playing by the river, and I watched her walk past; nobody took any notice of her. Eventually age caught up with her, and rather than let nature take its course, we tranquillized her and put her into a "granny cottage" enclosure where she could look out over the river and see her troop. During the first two days, when she spoke to her troop, the baboons challenged her, and many of the females charged her enclosure, sending her screaming into her sleeping quarters. After two days, she learned that she was safe, that nobody could hurt her, and she munched her bananas contentedly in full view of her troop, still offering friendly greeting grunts and looking dreamily over the river.

In December of 2007, two days before Christmas, Tripsy left us. We found her partially paralyzed and euthanized her. Did I do the right thing by prolonging her life? I cannot know; I can only do what my human brain tells me is right. She was my friend, and there is another empty space in my life.

Giovanni was a fine example of an alpha male. The troop lived well under his rule because he was wise and just and did not fly into tempers. The well-being of baboon troops depends on their leader's personality. If the alpha male is bad tempered and easily offended, he will dish out punishment at the slightest provocation. Whomever he abuses does not fight back but passes his or her anger and frustration onto the next-lower-ranked baboon, and so on, so that one small fight ripples through the whole troop until it reaches the lowest member. A bad-tempered leader is constantly surrounded by squabbling baboons. By the time one reprimand has been passed down through the ranks, he will have started another fight, which will also travel through the whole troop. The permanent members of the troop—the mothers and their babies—just have to live through the reign of such a tyrant, and luckily it does not last forever.

Poaching has always been a problem in Africa and will remain as long as people are poor and hungry. The area where our little farm is located was eventually declared a nature reserve, prohibiting any form of hunting. Snaring animals in this area carries heavy fines. It took a while for nearby landowners to accept our farm as a nature reserve. One of our reluctant neighbors took advantage of the fact that wildlife became more trusting after shooting was prohibited. He owned a butchery and encouraged Indigenous people to set snares, then sell him the carcasses. Many wild animals paid with their lives for the greed of this one individual before he was caught.

Giovanni was one of his victims. Late one morning, when the Longtit troop arrived, I found Giovanni sitting at my kitchen door, holding his arm out to me. There it was: a snare around his left arm, made of thick fencing wire. He had managed to break the wire, and, looking at the thickness of it, I marveled at his strength. I was all alone that day; even my faithful old Bennett had gone to do his monthly shopping.

I needed to dart Giovanni to remove the snare, but how would I be able to follow him before he went down? It takes a good two to three minutes before the drug becomes effective, and at my age, any baboon could easily outrun me. Again I resorted to one of my tricks. I lured Giovanni toward our feed room, which is enclosed. I am always utterly amazed at the almost-human understanding baboons have. Gently and calmly, Giovanni followed me, hopping along on three legs. Once I opened the gate, it was a piece of cake—he calmly walked in and helped himself to the food. I kept talking to him, telling him that he should not panic being locked up; then I slipped out and ran for the dart gun. When I returned, he was still peacefully eating sweet potatoes.

Baboons run when someone points an object at them, whether it is a gun or a camera or a stick. Baboons also try to keep a tree trunk, a rock, tall grass, or bushes between them and anyone who is observing them—just in case. And here I had to stand in front of a grown-up, injured male baboon, aim the dart gun at him, and shoot. How would he react? I spoke softly, explaining exactly what I had to do. Giovanni sat quietly, looking at me and chewing his food. Very carefully I aimed the dart gun at him; I could not afford to miss—and I knew I was the world's worst shot. Still no reaction from Giovanni: no protest, aggression, or attempt to hide. The dart went off and found its target perfectly. Giovanni hardly noticed being hit, carried on munching, and then slowly rolled over, mercifully asleep. I was elated, unaware that my nightmare was only beginning.

The wire snare had cut deeply into his arm. Nothing on earth could move it! I tried and tried, crying—almost praying—to have the strength to help him. I topped up the tranquillizer to keep him under sedation, then rushed off to phone for help. Luckily I contacted our local ranger, who arrived quickly. However, he also struggled to cut and remove the snare. Giovanni slept through the whole ordeal. Afterward I covered him with a blanket and let him rest until morning. When I checked on him the next day, he was sitting quietly. He had helped himself to some food and was watching me. Still, he showed no aggression or protest at being locked in. I opened the gate wide; he acknowledged it but took his time to hobble out.

We had examined the arm and were sure it was not fractured. He could not yet use the arm, and because he was so disabled, he made sure that the troop did not see him. He hid close to the feed room, and every time he saw me coming, he came close to ask for food, which I gladly gave, hoping that he would recover the use of his arm. But one morning no Giovanni was waiting. After a very thorough search of our area, we eventually found what was left of him. Judging from all the signs, he had put up a desperate fight. The grass was flattened around a big tree, blood was everywhere, and only his head and limbs remained. Perhaps, due to his aching arm, Giovanni had become careless in climbing his sleeping tree and so was an easy target for a hunting leopard. He must have fought back bravely, but the odds were against him. He died like most of his kind, battling the statistics of survival, jeopardized by human greed and selfishness.

Giovanni's memory has the aura of a truly worthy leader, an amazing understanding of what it means to be handicapped, and a genuine appreciation for one small and aging person's attempt to help. He headed the troop for almost four years, and it was a time of bliss and peace.

My fight for baboons in South Africa has been a long and hard road and is far from over. I have won many battles, and many I have lost, but baboons have become the purpose of my life. They are my children—my family—and I love them very dearly. They have touched my heart more than any human ever could.

Acknowledgment

Many thanks to the editor for turning my notes into a complete essay.

14

—

A Veterinarian with Conviction

Karmele Llano Sanchez

When I was very young, I knew I wanted to be a veterinarian. My interest in animals continued to grow as I got older, and I soon realized that I could not be a common vet—working in a small animal practice for profit. My expectations and goals required more: I wanted my life's work to contribute to conservation.

While working in Venezuela at the Asociación de Rescate de Fauna Amenazada (ARFA), a rescue center for wildlife, I decided to focus my veterinary work on nonhuman primates. So, when I received my bachelor's degree in veterinary medicine, I headed for Stichting AAP (an ape foundation in the Netherlands, http://www.aap.nl/index.php) to work with primates who had been rescued from very difficult situations. AAP was founded in 1972 as a way station for exotic pets who needed permanent sanctuary. In 1996 the foundation moved to a larger place in Almere and eventually became the premier primate sanctuary in the European Union. In 2007 it housed 274 mammals, including 179 primates, most of whom came from Rijswijk's Biomedical Primate Research Centre.

As I worked with primates in the Netherlands, I realized that I wanted to complete the cycle—to be able to release these beleaguered individuals back into the wild as I had done in Venezuela. This could only happen in a country with the proper habitat, so I set my sights on Indonesia for 2003. From Spain to Venezuela to the Netherlands—cards were falling into place. I was bound for Indonesia, where I would eventually found an organization specializing in lorises and macaques.

Indonesia is one of the most biodiverse countries in the world, and it also has one of the largest numbers of primate species, including various lorises, leaf monkeys, macaques, orangutans, gibbons, and others. Macaques and lorises are the most neglected of Indonesian primates, so I and my friends in Indonesia decided to set up a rescue-rehabilitation center and sanctuary specifically for these two species.

In Indonesia macaques are considered pests like rats! Consequently, they are mistreated and neglected. Indonesian macaques are often sold in horrible bird markets—terrifying places. It takes great courage for me to enter an Indonesian bird market, where I know I will be confronted with wild-caught, caged animals suffering from pain, depression, sickness, and terror. Hunted wild animals are taken to these markets to be displayed and sold; while they are there, people and other animals surround them, which is frightening and very stressful. The air feels dense and full of pathogens in these bird markets. Wild animals under this kind of stress are immunodepressed, and contact with new pathogens in crowded markets makes them even more susceptible to illnesses. Mortality is high, but the exotic pet trade is very profitable. Some people in Western countries—those who like to keep exotic pets—do not realize how much these animals suffer before—and even after—they finally find their way into a human home. These markets are common throughout Asia, and if there is a hell for animals, it must be very similar to these places (or to animal laboratories in research facilities).

Little baby macaques are always for sale in Indonesian bird markets. At this age, they are cute: they have small baby hands and feet and beautiful round eyes that differ in color from brown to green and yellow. Their hair is gray or brown to golden, and sometimes the long-tailed macaques (*Macaca fascicularis*) have funny-looking Mohawk hairstyles. The pig-tailed macaque (*Macaca nemestrina*), with a comical piglike tail (as its name suggests), is darker brown. Both of these primate species are so cheap in these markets that almost anyone can afford to buy one. Sometimes these baby monkeys are loved when they are cute and cuddly. Still, they are purchased as status symbols, and people tend to display them in front of their houses in small cramped cages or chain them to power poles or trees alongside roads, without a roof or any shelter from the hot sun or heavy rain. They are left there with no food and no life. As they reach puberty, their novelty and cuteness disappear, and they usually become aggressive. They break their chains or cages, and people use increasingly cruel methods to confine these displaced primates.

In the countryside, where humans encroach on primate habitat with palm plantations or simply because of human overpopulation, macaques tend to raid crops. Consequently, they—along with orangutans, who also enter palm groves, fleeing from dwindling forests and seeking food—are officially labeled as pests and killing these intruders is therefore justified.

Macaques are also exploited as an export commodity. They can be farmed, in which case they are kept in unconscionable confinement in tiny cages, then sent overseas. They are often sent to China, where they are used for medical experimentation, or to Western countries for biowarfare tests. Needless to say, no one would wish to be sold into a life of such exploitation and suffering. The Chinese people also eat macaques, cracking open their skulls while they are still alive and fully conscious and eating their brains. The Chinese consider this dish a delicacy.

With this in mind, I set my sights on a rehabilitation center where at least some of these unfortunate primates might find sanctuary or be rehabilitated and returned to the wild. Luckily I was not alone entering this monumental task; I teamed up with a committed, motivated group of people in Indonesia who had previously worked at a rescue center for wildlife—Tegal Alur—in Jakarta. We all agreed that we needed funding, so I returned to Europe, and while searching for funds, I came across Alan Knight, the chief executive of International Animal Rescue (IAR). It was not until several years later, however, that Knight and his team at IAR agreed to help us set up a rescue-rehabilitation facility in Indonesia. It wasn't that IAR had not wanted to help us all along, but funds are short, and need is great. IAR's initial donation funded our first release—a group of macaques who are now romping in the forests of Lampung, Sumatra. We renamed our fledgling organization IAR Indonesia in appreciation for their help.

Funds from IAR also enabled us to build a rehabilitation center for macaques and lorises near Bogor on the island of Java—the first Indonesian rescue and rehabilitation facility for these needy and neglected species. With the help of IAR, we created facilities where, for the first time, macaques, lorises, and other primates—as well as our rescued dogs and cats—can escape the otherwise-eternal misery of confinement in human hands and instead be rehabilitated and return to their natural way of life.

The annual operational cost of our rescue-rehabilitation center (run by Yayasan IAR Indonesia) is about $150–200,000. Most of this cost is covered by IAR in the United Kingdom, which is largely funded through private donations from individuals in Western countries, including legacies. However, this figure does not include many of our programs, such as forest protection and surveys, community development, law enforcement, awareness campaigns, and post-reintroduction monitoring of rehabilitated animals. These projects are mainly financed by grants from organizations or foundations, or even government bodies. We also receive help from other generous organizations that are funded by private donations, such as the International Primate Protection League (IPPL). Every single dollar counts. IAR Indonesia could not operate without this assistance. I feel eternally grateful to those who send money to help us help the animals. The trust of generous donors creates a sense of urgent responsibility, and we are very careful with every cent we spend.

As an activist, I admire people who are dedicated to fund-raising to help animals. While working with primates is challenging—and sometimes depressing when we see animals wounded needlessly or watch them die helplessly—at least we have the chance to see, and sometimes even know, the animals we help. Fund-raisers are perhaps the most invisible of activists, yet they are fundamental to almost everything that we do. All other branches of rescue and rehabilitation depend on fund-raisers. Without them—and without their funds—we are finished.

We also have many volunteers from overseas who come to help, and we appreciate them tremendously. We also hope that they will carry their new knowledge about primates—and animals in general—back to their lives at home and their communities. Volunteers are part of the way that we reach people, telling them that we must act quickly if we are going to save the world's remaining primates, or that we must rethink animal experimentation because it is cruel and increasingly unnecessary, or that wild animals should never be kept as if they were pets. Please know that you can help primates—and all animals—from home by spreading the word and raising awareness or fund-raising and offering a donation.

IAR Indonesia's main goal is the rescue and rehabilitation of primates, but we also investigate illegal trade in wild animals and run education and awareness programs and conservation projects that include forest restoration and community development. IAR Indonesia works with both animal-welfare and environmental organizations, two advocacy groups that, unfortunately, too often fail to support one another. We are proud to be a bridge between these two advocacy approaches.

IAR Indonesia also rescues other needy primates, especially lorises, who are critically endangered. Many of these little primates are exploited for the wildlife pet trade; mafias trade and traffic these primates because they attract customers with their huge, beautiful eyes. Such trafficking has depleted entire populations of lorises, who are then sold in horrid bird markets and alongside roadways. They are exported as far as the Middle East and find many homes in Japan, where they are very popular. If you have friends in Japan, please let them know that lorises belong in the wilds of Indonesia, not in living rooms.

Often people who have pet lorises do not even know what species they have purchased. Consequently, these unfortunate animals are fed bananas and papayas instead of their natural diet: insects. This unnatural diet causes disorders such as diabetes and osteodystrophy (lack of calcium), further increasing mortality in captive lorises.

There are three known species of slow lorises in Indonesia: *Nycticebus javanicus* from Java island, *Nycticebus coucang* in Sumatra, and *Nycticebus menagensis* from Kalimantan in Borneo. They are normally extremely slow (hence their name), but they can also be very quick and react with impressive speed when hunting or trying to escape predators. They hunt insects but also sometimes eat small birds, mammals, and reptiles. Despite their diminutive size (not more than one kilogram), they are excellent hunters. They move slowly and quietly toward their prey, then simply grab victims with their hands.

Lorises are the only venomous primates, which makes them even less suited to be treated as pets. Their venom gland is located in their arms. Mother lorises mix this venom with their saliva and cover their babies with this toxic substance before they park their helpless young so they can go hunting. They also use their venom to stun prey. Loris bites are painful. They have small canine teeth, but when they suck venom from their arms and mix it with saliva, the poison that enters a human victim's bloodstream can cause an anaphylactic reaction. Consequently, traders cut loris teeth with nail clippers or wire cutters before they sell them, which not only causes a great deal of pain but often results in deadly infections.

Lorises are called malu-malu in Indonesian, which means "timid." They are indeed timid, if not evasive. They are one of only a few nocturnal primates, which makes them particularly ill suited to captivity since they sleep all day. Furthermore, they do not like to be watched or touched, and captive lorises suffer from stress and diseases; their mortality rates are high. But lorises are cute; they have big eyes that resemble characters in Japanese cartoons. Perhaps this is what makes them so popular as pets, along with the fact that their hands resemble those of small humans. But humans also seem fascinated by the unique features of lorises, for example, their venom and nocturnal ways; lorises also have a long, odd-looking grooming claw in one of their toes.

Local legends hold that these small primates are either a symbol of luck or the opposite. Many Indonesians believe that lorises have magical powers, and they are willing to torture and even kill lorises to gain access to these magical powers. For example, in some areas of Indonesia, lorises are buried under asphalt in the belief that this will prevent highway accidents. In other areas, they are burned alive, then squeezed to obtain a love potion called minyak kukang. Some people believe that this elixir—when offered to a beloved—will produce eternal love. Still others believe that burying a loris on someone else's property will magically cause that person to sell this land cheaply to the one who buried the loris. The truth or falsity of such claims seems irrelevant considering the suffering and premature deaths these beliefs cause.

Very little is known about these evasive creatures, partly because studying a shy, nocturnal animal in dense forests is very difficult. Some studies have shown that tree gum is a very important source of carbohydrates and energy for lorises, so they are sometimes called gummivores. Lorises use their special comb-shaped incisors to break open the tree, then suck out the gum. Other studies are uncovering completely new information. For example, despite the fact that they have always been considered solitary, new research has shown that these little primates are indeed very social: they sometimes live in family groups, they embrace and kiss their partners, they nurture their young for a prolonged period of time, and they make both friends and enemies in their social groups—just like we do. Enemies can deliver nasty bites, but some of the lorises we have socialized become friends at first sight and sometimes even partners in raising families.

As a veterinarian with IAR Indonesia, I can relieve some of the suffering that lorises experience because of human trafficking. In our clinic in Ciapus, Java, we radiograph teeth, operate on mouths, and remove the many broken teeth and tooth fragments from inside these primates' gums, relieving a great deal of pain. I learned this dental operation from my friends, Dr. Paolo Martelli and Dr. Karthiyani Krishnasamy, two vets from Hong Kong who often come to help at our center. I am grateful to these two people, and others like them, for helping me to help so many nonhumans.

After removing ten to twenty teeth from just one loris, it is amazing to see these resilient little primates happily eat, when previously every meal was painful. For example, Jane's mouth was particularly painful. All of her teeth had been cut; I could only save three of her molars—all the rest had to be removed. Nonetheless—and much to my surprise—after her operation, she started eating her food with gusto alongside the other lorises at IAR.

IAR Indonesia's goal is to release lorises back into the wild but only if they still have their teeth, which is rare. Thanks to the recent introduction of radio collars, lorises who have lost some of their teeth will also have the chance to be released. Further research is needed to find out whether these individuals can cope without teeth in the wild, and IAR is poised to take a primary position in this much-needed research.

While those of us at IAR specialize in lorises and macaques, we sometimes help other animals. Recently we were able to help a wild Javan leopard, who had been caught and severely injured by a wire snare. We were afraid that he would die. Much to our amazement, he recovered completely, and we were able to return him to the wild.

We also help other primates—such as gibbons—who are common victims of Indonesia's illegal pet trade. Like macaques, they are sold as cute babies in spite of the fact that gibbons normally live in family groups where babies stay with their mothers until they are five. Baby gibbons are extremely dependent on their mothers, clinging to them for the first years of their lives. Gibbons are also one of very few monogamous species, staying with their partners for a lifetime. To catch these cute gibbon babies, hunters kill the mothers and sometimes the entire family, including father, brothers, and sisters. Baby gibbons sold into human confinement suffer greatly and are often depressed and traumatized because of the loss of their family. Some of the gibbons that I have treated at rescue centers were so depressed that they simply stopped eating.

Saar was a typical small baby gibbon found by forestry officials. I was given the task of taking care of him. He was just the size of my hand and probably five or six months old when he arrived. Saar was a Javan gibbon—also called silvery gibbons because of their beautiful gray-silver fur—with a black face and hands. Gibbons have very long arms, longer than their legs. They use them to swing from branch to branch. Saar was so beautiful and so small that he seemed to be only eyes and arms. He clung to me the way he would have held onto his natural mother, but after some time, he slowly gained a measure of independence and was socialized with another gibbon in a rehabilitation center.

When Indonesia's Nyaru Menteng Orangutan Reintroduction Project in central Kalimantan in Borneo was overwhelmed with an influx of new primates, the staff requested my help. The palm industry is decimating orangutan habitat in Indonesia, leaving these beautiful beings to die and/or be evicted from their homes with nowhere to go. The number of orangutans arriving at the project has been increasing steadily with the growth of the palm industry. At that time, they had more than four hundred; now they have almost seven hundred. Please take care not to buy any products containing palm oil—neither for your kitchen nor personal use!

At the Orangutan Reintroduction Project, I had the privilege of working with Lone Drøscher Nielsen, whom I quickly came to admire. She is extremely brave and independent, hardworking and intelligent, and she has also given up absolutely everything in her life to help orangutans. She deserves the respect and admiration of all who care about nonhumans.

September 21, 2005, the day that I arrived at Nyaru Menteng, another visitor also came—a baby orangutan. His name was Bintang, which means "star" in Indonesian. He was very skinny. His bold face and head were a bit darker than usual. Healthy orangutans have shiny orange-to-red fur, which grows longer as they age, but Bintang's coat did not look very shiny. His eyes were watery, the same as our eyes when we get the flu. He had been kept on a wooden platform—just two by two meters—which hung over a river and was next to a chicken cage. Orangutans never go into the water, so he was confined to this small platform day after day.

Not surprisingly, when Bintang arrived, he was very sick with a high fever. Consequently, he was the first orangutan to spend the night with me. Of course, I hardly slept because I was worried about Bintang; I kept waking up to make sure that he was still breathing. For two months, his fever raged until he finally succumbed. We figured out why he was sick just before he died: tuberculosis. The last night that I visited him in the baby schoolroom, he was beginning to show signs of neurological damage. It is likely that mycobacteria, the pathogen that causes tuberculosis, had invaded his brain. He struggled to breathe, and Lone and I stayed with him all night; we were with Bintang when he breathed his last. We could do nothing but offer him companionship in his last moments; the ill effects of human negligence were more than his little body could handle.

All primate species share a number of diseases, but tuberculosis is a bacterial infection among humans, one that can be transmitted to other primates. Watching little Bintang die was very painful, and it helped me see just how much suffering we humans inflict on animals because of our egoistic desire to have wild animals as pets. Bintang would never have contracted tuberculosis if he had been left in the forest with his mother, where he belonged. Had Bintang not been stolen, sold as a pet, and imprisoned outside someone's home, he might have lived a long and happy life in the wilds of Indonesia.

It was also at Nyaru Menteng that I met Yetno, a big cheek-padded wild orangutan. When male orangutans reach adulthood, they develop wing-shaped cheek pads on their faces. He was big. Though they are hefty, practically the only source of meat in their diet is insects. They are harmless and would never attack except in self-defense—for the sake of survival.

Yetno had been rescued from a palm-oil plantation and arrived in terrible condition. He should have been so strong, yet he was miserably sick and could only lie helplessly in his cage. He looked at me with desperate eyes. As one who heals, I wanted so much to save him, but the ill effects of humans again proved too much. His autopsy revealed a cranial fracture. Plantation workers had hit him in the head, breaking his skull. His sad and desperate look seemed to ask, "Why?" I do not have an answer. Do you?

Palm-oil plantations destroy natural habitat and unique primary forests; they are the main threat to biodiversity in Indonesia and to tropical rain forests. These plantations deplete reserves of peat forest, release enormous amounts of carbon into the atmosphere, and cause a lot of suffering in the rain forests of Kalimantan. Many nonhumans are lost to the wildlife pet trade, but the greater threat to species and individuals comes from loss of habitat due to palm plantations. When their environment is destroyed, nonhumans lose both their homes and their sustenance. They are also burned alive in forest fires that are set to clear land for plantations. Hunters and plantation owners brutally kill them when they no longer have the protection of the rain forest, and they also become easy prey for people who trade in exotic animals. The growing palm industry—and our Western world's incessant demand for palm oil—are destroying Borneo's wildlife. If we do not do something soon to protect these forests and their inhabitants, most of Indonesia's nonhuman primates will disappear into extinction. Yetno, who would otherwise have been a powerful and athletic individual, springing from trees and eating fruits and leaves, died because of the palm industry.

I have been working with primates as a veterinarian for seven years. Even before I turned my focus to primates, I considered all animals to be morally relevant and my equals. Neither humans nor the other great apes should be placed at the top of any species hierarchy. Orangutans are similar to humans, as are our closest cousins, the chimpanzees. However, this does not make orangutans or chimpanzees cooler or offer them special moral status. There is no reason why caring for apes should be more important than working with or for cattle or cats, chickens or rabbits. Every individual who can feel pain—whether physical or emotional—deserves equal respect and protection.

15

Volunteering in Thailand

The Gibbon Rehabilitation Project

Fiona Mikowski

It all began when my girlfriends and I decided to take a short vacation to Thailand. I was halfway through my bachelor of animal science degree and needed a break from all the books to figure out what direction I wanted to go. I came across the Gibbon Rehabilitation Project (GRP) in Phuket, Thailand, and I thought to myself, "What a perfect project it would be to help with nonhuman primate rehabilitation!"

Living at the GRP, which is located in a small village some distance from the main town of Phuket, I was able to see more of the way the Thai people live. Some volunteers have a hard time adjusting to Thai village life, which consists of a few small shops and a fruit and vegetable market twice a week. Not all the people in the village agree that the GRP is a good idea, but they all accept it and understand that it brings volunteers and other tourists to their village—an important economic benefit. The villagers are both Buddhists and Muslims. There is a mosque near the project, and we could hear the call to prayer several times a day; Buddhist temples surround the GRP.

The Thai people are very kind and make visitors feel welcome and at home; they are always eager to make new friends. Most of the GRP staff members have never been out of Thailand, so they are fascinated by our ways, especially our tendency to complain. Thai people have very little money and few belongings, yet they are always smiling and friendly. Thai staff members at the GRP make a minimal wage, and their pay is not even the same each month since the project subsists on donations and the gibbons' needs must come first. The workers could look for a job elsewhere and make more money, but they choose to work for the GRP because of their passion and respect for gibbons.

The GRP was established in 1992, when it became illegal to take a gibbon from the wild. Sadly, gibbons are still stolen from the rain forest they call home because the illegal trade remains profitable. Gibbons are valuable and can be sold to entrepreneurs as tourist attractions or to individuals as pets. The GRP specializes in tending white-handed gibbons (*Hylobates lar*), who are native to Southeast Asia and are included on the Red List of the International Union for Conservation of Nature (IUCN) and designated as endangered in Appendix I of the Convention on International Trade in Endangered Species of Wild Fauna and Flora (CITES). When I arrived, I knew nothing about gibbons or the threats these amazing primates face, but I was eager to learn.

And I did. I learned that gibbons were poached to extinction on Phuket Island more than twenty-five years ago. The GRP aims to repopulate the last remaining rain forest in Phuket, the Khao Pra Theaw Non-Hunting Area, by rehabilitating and releasing white-handed gibbons. They also educate local communities and tourists about the importance of conserving these individuals and their habitat.

I discovered that gibbons live in monogamous family groups—one male to one female—with up to three offspring. They are very territorial and rarely migrate to other parts of the forest, even when their habitat is disturbed. I also learned that this practice makes gibbons vulnerable when hotels, shrimp farms, fruit trees, rubber-tree plantations, or hunting—poaching for the lucrative exotic pet trade—destroy their forest homes.

On the morning of my first day at the GRP, a most unusual sound awakened me, a sound like nothing I have ever heard before. I can only describe it as some sort of sweet siren. I was perplexed by this sound as I headed to the office to begin my first day of orientation.

I was shown around the office, quarantine site, and the Center for Conservation, Education, and Fund-Raising. I watched a thirty-minute documentary on the project, and through this video I learned what I had heard so early in the morning—the voices of gibbons. Our rooms were right next to the quarantine site, so their calls were very loud.

Gibbons are placed in the quarantine area when they first arrive. Here they are given full medical examinations, undergo deworming, and receive tuberculosis and blood tests. The blood samples reveal whether the gibbons have diseases such as herpes simplex virus, hepatitis A, hepatitis B, or HIV, which they may have acquired from needles or being given tourist's drinks to share. If gibbons test positive for any disease, they cannot be released back into the wild. Nor can these individuals be released if they have any serious body deformations. In such cases, these gibbons remain in quarantine or rehabilitation, teaching visitors what happens to gibbons when they are forced to be pets or tourist attractions.

Even if newly arrived gibbons are free of disease, they must stay in quarantine for at least three months. For some of these individuals, this is the first time they have encountered or heard another gibbon. They have much to learn about being gibbons, and only after they begin to act more naturally can they be transferred to the rehabilitation site, where they undergo an extensive rehabilitation program.

On my second day, I was in front of the office by 6:30 a.m., eager to start my work at the rehabilitation site, located just a short drive away. The rehabilitation site houses healthy primates in an environment where they can learn natural gibbon behavior. They practice brachiating, arboreal locomotion where they use their arms to swing from tree to tree. In rehabilitation they swing from the sides or top of their large cages and from several large bamboo sticks to strengthen their arm muscles in preparation for life in the wild. They eat natural foods and have minimal contact with humans.

At the rehabilitation site, gibbons are weaned from human dependency. Juvenile gibbons share cages, while adults are placed individually in cages close to others. This gives them the opportunity to form pairs without conflict. The gibbons are watched carefully, and if two adults show interest in one another, they may be placed in the same cage. Only gibbons who have mated and created strong family bonds are released into the wild. If they adjust to social life, the gibbons graduate into larger enclosures farther from human eyes, thus decreasing their daily contact with people.

I often began my work at the rehabilitation site by separating fruit and vegetables for the upper and lower gibbon groups: vegetables in the morning and fruit in the afternoon. The vegetables were washed and cut into even pieces, and gibbon balls were prepared, which consist of rice and a few treats that the gibbons are especially fond of. As the rice cooked, the vegetables were distributed; one person fed the upper group, and the other fed the lower one. We also filled their water bottles.

At 8:00 a.m., we were served breakfast at restaurants in adjoining Khao Pra Theaw Non-Hunting Area. At 9:00 a.m., we divided the work that needed to be done that day among all of the volunteers. Each day we cleaned feces and food scraps from underneath the cages, made sure the cages were intact, did behavioral observations and health checks, and of course created and delivered gibbon balls. Each day we also took on an extra job, whether sweeping pathways, cleaning water bottles, or cutting banana leaves for the gibbons—another treat.

One of the gibbons, Endoo, also required extra attention. Because of human neglect and abuse, she self-mutilated, mostly biting and scratching her arms until they bled. We provided her with enrichments to keep her occupied and decrease her chances of self-harm. We sprayed a scented fragrance into her cage, offered her a large ice cube with syrup fruit and nuts inside, and provided a maze box filled with holes that we stuffed with food, all of which engaged her mind and hands. She especially enjoyed the box of treats and spent a lot of time pushing the food around in the maze until it fell out of a hole into her hands.

Lunch followed from 11:30 to 12:30, after which we washed and divided the fruit for gibbon lunches. Once lunch had been distributed, we cleaned the kitchen and were picked up at 2:00 p.m. and taken back to the office. We then had free time for the rest of the day.

Sometimes I worked at the tourist desk, also known as the Center for Conservation, Education, and Fund-Raising. Like the rehabilitation site, this building is located near the entrance to the Khao Pra Theaw Non-Hunting Area, near Bang Pae waterfall, which many tourists come to see. At the center, volunteers greet tourists from 9:00 a.m. to 4:00 p.m., explaining the project's aims, why gibbons come to the GRP, and what people can do to stop gibbon poaching. They emphasize the importance of conserving habitat and suggest that people avoid supporting businesses that exploit wild animals. In addition, these volunteers escort visitors to a viewing area, where they can see GRP gibbons. They describe the gibbons' individual histories, detailing what they endured before arriving at the GRP. Since the GRP is funded largely by donations, volunteers encourage visitors to support the project by adopting a gibbon, purchasing merchandise, or simply leaving a donation.

At the GRP, each volunteer is given the chance to accompany one of the staff members to the forest to gain a richer understanding of gibbon habitat. An extremely exhausting half-an-hour walk through Khao Pra Theaw leads to the training cage. This is the first stage for gibbons leaving the rehabilitation site on their journey to forest freedom. This large cage allows a family of gibbons to explore and adapt to the rain forest environment for about three months. They are then transferred into the acclimatization cage.

The acclimatization cage is suspended twenty meters above the forest floor. The cage door stays shut for at least ten days, allowing the gibbon family to adapt to the sounds and smells of their new surroundings without intrusion. Then the door is opened, allowing the gibbons to move freely between the forest and the cage. As they become more confident in the forest, the cage door is closed, encouraging the gibbons to establish their own territory. The GRP continues to feed these now-wild gibbons for a while, but the food gradually decreases as they become more capable of foraging.

The day that I visited the forest, the Payu group—consisting of a mother, father, juvenile, and baby—was in the training cage. We checked to be sure they looked healthy and offered food, using a system designed to prevent human contact. We placed their food on top of their cage, encouraging the gibbons to stay in the upper canopy of the forest, away from predators.

We continued into the forest until we reached the feeding basket, which we filled. This basket was for a gibbon family that had already been released, the Arun group, which had not yet been completely weaned from human handouts. These provisions not only assure survival but also allow staff to keep an eye on new releases.

We then traveled into the Hope group territory, an area established by the first group of gibbons released by the GRP. On that particular day, we did not spot any of the Hope gibbons. We did not even hear them calling, but their calls are usually recorded daily to allow staff to keep track of these recently rehabilitated gibbons.

At the end of my three-month stay, I heard that the Payu group would soon be released. I decided to extend my time at the GRP to see this gibbon family released into the wild. Releasing a gibbon family is a very special occasion, and the GRP involves the local villages and schools, making an afternoon event of every release. For the Payu release, many schoolchildren participated in a skit and donated artwork, which was sold to support the GRP. Although most of the festivities were in Thai, volunteers received a special T-shirt and were included as much as possible, given that we were language challenged. I could see that the Thai staff had successfully educated local community members about the problems facing gibbons and the importance of the GRP. Even little children seemed to understand that poaching is cruel.

Eventually a group of staff and volunteers left the festivities and headed for the acclimatization cage, where the Payu group had been transferred a few weeks earlier. As we made ourselves comfortable in a spot where the gibbons could not see us, the staff opened the cage. The mother and baby quickly fled and dashed up to the highest area in the trees, where they were clearly comfortable and felt quite at home. It was a beautiful moment.

The father and juvenile did not want to leave and stayed in the safety of their cage. The mother and baby would not be able to survive in the wild without the protection of the father. We waited all day, but the father did not choose to leave, so the whole family was put back into the cage to offer more time for them to adjust to their new surroundings.

I kept in touch with friends at the GRP and was later informed that the Payu group was released in a different area, farther from the Arun group's territory. This time, when the cage door was opened, the whole family moved out into the forest canopy. Staff and volunteers continue to observe the Payu group; it appears that they are adjusting well to life in the wilds of Thailand.

The GRP has not always had successful releases: some gibbons have died, while poachers have killed others. Every release is a learning experience, and the GRP constantly works to improve each rehabilitation step in spite of limited funds.

GRP Thai staff members are hard working and dedicated to the gibbon cause. Those who wish to volunteer must have passion for the work and respect the staff and other volunteers. GRP staff works from early morning until late at night, then sleeps among the volunteers in rustic accommodations, sharing a bathroom, kitchen, and lounge area. I have never seen such hardworking people who have so little—yet they never complain.

GRP Thai staff members often speak in their own language, sometimes making it difficult for volunteers to understand what is happening. Luckily the GRP holds volunteer/staff meetings once a week to inform everyone of the overall plan, the current conditions of each gibbon, and any other relevant matters. In all situations, volunteers need to be patient and understanding: for many Thai staff members, the GRP is their life.

As time went by, I realized that each gibbon at the GRP had a completely different face and personality. Their expressions are amazingly similar to ours. Two gibbons in particular captured my heart during my stay as a volunteer: Bo and Tam.

Tam was a blonde female who had been kept as if she were a pet. As she matured, she began to act like a territorial gibbon and tried to bite her owner's child. Her owner beat her so savagely that she had to have a hand and foot amputated. She was then passed to different owners, one of whom put her in a cage with many other gibbons without any introduction. Since gibbons are very territorial, the others attacked her, and she lost all but one finger on her remaining hand. Finally, this poor gibbon was brought to the GRP. Because she only had one finger, she couldn't groom herself properly; ticks attached themselves around her eyes, and the staff had to remove them. The skin between the last remaining finger and her finger stumps had a tendency to get too moist, which also caused problems; each day a volunteer powdered her fingers, which she thoroughly enjoyed. Despite all Tam's human-caused pain and suffering, she was gentle and friendly and had somehow forgiven us.

Tam's cage was next to Bo's in the rehabilitation site. Bo was born in the wild like the majority of gibbons at the project. He was probably used as a pet or business prop for tourist photos because he arrived at the GRP emaciated, with his milk teeth filed down. He was in very bad condition. Nonetheless, after some time at the GRP, he fathered offspring with a gibbon named Lek. They were all released into the wild in 2003.

Unfortunately, it soon became apparent that Bo had been too institutionalized to be released. He abandoned his family in the forest and returned to the project six times. Since he did not want to stay in the forest, the GRP staff decided that he should be kept at the rehabilitation site. I remember him as a very sweet, gentle gibbon who was always looking for attention from volunteers. When you looked into his eyes, you couldn't help falling for his loving charm.

Since both Tam and Bo had become permanent residents, the GRP staff needed to sterilize Bo: though there is great need for wild gibbons, there is no need for more captive ones. Putting these two damaged individuals into one cage was a long process. Volunteers conducted one-hour observations three times a day over the course of several weeks to keep an eye on the developing relationship between Bo and Tam. They groomed each other through their wire doors, and Bo offered Tam a banana. Staff had to be very careful since Tam had been through such horrific experiences with other gibbons. The door between their cages was eventually opened. After a couple of weeks, Tam began acting differently and seemed stressed, so they were returned to separate cages. They continued to groom each other and play through the wire—which is much better than living completely alone—but they remain permanently housed in separate quarters.

One other gibbon who touched my heart was Joy, one of the oldest gibbons at the GRP. Joy had been captured from her parents and sold into slavery as a pet. As she matured, she became territorial and aggressive—as is natural—so her owners locked her in a small cage in a dark basement and forgot about her, feeding her only once or twice a week. This caused Joy to become very agitated at feeding times, and it took years at the GRP for her to improve. Since she is an older gibbon, there is a chance she will never completely overcome this behavior, so she is not a candidate for release, which is always heartbreaking. Permanently caged gibbons are victims of human greed, ignorance, and indifference.

Before arriving at the GRP—during my vacation with my friends—we came across several baby gibbons being exploited as props for tourist photos. Like most tourists, I had no idea that this practice was illegal, nor did I understand the suffering these small businesses cause until I began volunteering at the GRP.

Gibbons are illegally kept as pets and tourist attractions in Thailand. The only way to capture young gibbons for these purposes is to take them from the wild, shooting the father and mother to whom the baby clings. If the baby avoids being shot and survives the fall from the upper canopy, he or she is caught and sold. These violently orphaned babies are kept as pets or trained to pose in pictures for tourists on the beaches or streets; tourists unknowingly support this illegal and damaging industry—the trade in exotic animals—when they pay to have their photographs taken with a gibbon or any other wild animal. Alternatively, businesses such as bars exploit these primate victims as gimmicks to attract patrons, sometimes dressing them in diapers or forcing them to smoke cigarettes or drink alcohol.

Gibbons reach sexual maturity at six or seven years of age. At this point, they develop territorial behavior and large canine teeth, and they become quite unpredictable. If they are held as pets or used as tourist attractions, their owners will probably file down their canine teeth or pull them out completely. They may also drug the maturing gibbons to keep them calm and/or awake until late at night, when gibbons are normally asleep, to please their patrons. Other owners don't know what to do with gibbons who have reached sexual maturity because they become too much of a risk and are useless as a pet or prop in photographs, so they chain or cage these individuals for years, often causing permanent psychological damage.

Some of these gibbons are lucky: they are confiscated by the police and/or given to the GRP. This gives them a second chance at life in the wild, if possible. Otherwise they can live in a spacious, natural setting, where they will be well tended for the rest of their days. Unfortunately, the owners of these confiscated gibbons often acquire another baby, who is easier to handle and more attractive to tourists—at least for a while.

Please don't ever enter places that keep captive wild animals or pay to have your photograph taken with an exotic nonhuman. Don't even buy gibbons to rescue them because this supports the exploiter and allows him or her to buy another violently orphaned gibbon. This practice continues to reduce the number of gibbons who remain in the much-depleted rain forests.

All of the gibbons housed at the GRP once lived in the wild with their families, but humans destroyed their families and subjected them to a cruel, pointless life to satisfy human needs and desires. These unfortunate gibbons became slaves for profiteers. They have no idea why they are confined or forced to sit with tourists. When they mature, they are punished for exhibiting their natural behavior. Yet somehow—once they reach the GRP—these gibbons forgive and forget, and the project continually works to provide these misused individuals with the possibility of a second chance in the wild.

Some GRP gibbons have suffered so intensely that they never recover and must remain in cages all of their lives. We cannot blame the Thai people since it is Westerners who most often pay to have their photographs taken with gibbons or are drawn into pubs that have gibbons on show. Worldwide education is needed to end this horrible situation and ensure gibbon survival.

Though living in extreme humidity in the midst of an unforgiving rainy season thick with mosquitoes, cockroaches, frogs, and snakes, I loved my time at the GRP. When I first arrived, I was mystified by the sound of gibbons singing. Now—back home—I miss this sound terribly. We must all take responsibility to ensure that this remarkable music of the rain forest does not disappear forever.

Acknowledgment

Special thanks to Lisa Kemmerer for turning a very rough draft into a completed essay.

16

Friends Are the Family We Choose

Paula Muellner

I always thought that a good story needed two things: a meaningful or entertaining premise and a good delivery. I also felt that my life was full of events and thrills—I've trekked sky-high peaks, careened down dangerous rivers, sailed the eastern coast of the United States in a 131-foot schooner, and walked with wild chimpanzees in Africa. Yet I considered myself a horrible raconteur: I never had a good delivery.

But people love stories. They are a valuable way to learn, a way to make meaningful experiences out of the good and bad events in our lives.

Up until five years ago, I didn't know how to tell a good story. Perhaps it was because I spent too much time thinking about responsibilities, so I never focused on the small, simple pleasures that make life worth living. Things have changed now, and thus, my story begins.

In the latter half of 2003, I was offered an extraordinary position with Chimps Inc., a chimpanzee sanctuary in Bend, Oregon. It was difficult for some of my family and friends to understand why I wanted to leave a high-paying government job to work for a nonprofit sanctuary.

In retrospect I realize that my family and friends didn't understand what a chimpanzee sanctuary was. Most people believe that sanctuary workers simply play with monkeys all day long. But I was well aware of the purpose and need for chimpanzee sanctuaries. Thousands of monkeys and apes have been captured and bred to satisfy the medical industry, entertainment sector, and pet trade.

Chimpanzees, who share nearly 99 percent of our genetic makeup, have become unfortunate subjects for decades of biomedical research. For these nonhuman primates, a life in captivity is inherently cruel: they are separated from their mothers at birth and raised by unnatural parents, then their bodies are used for science—whatever damage or pain this causes is part of the procedure.

As young primates mature—whether in medical labs, entertainment venues, or the pet trade—they become rambunctious, aggressive, and unpredictable. These innate behaviors are harshly repressed—often physically. By the time young chimpanzees reach eight, human handlers usually consider them unmanageable. Consequently, these primates become victims of isolation, abuse, and neglect, which results in psychological damage.

Since chimpanzees can live to be sixty years old, the task of providing a suitable environment for these primates is incredibly difficult. Unfortunately, many of these apes are shuffled off to roadside zoos, breeding warehouses, or worse, they are silently euthanized. Only a small number of lucky primates are given the opportunity to live out their lives, unfettered, with others of their kind in sanctuaries like Chimps Inc.

And that is why I was so eager to work for Chimps Inc. I enthusiastically accepted the job offer because I knew how few positions like this exist. At the time, there were only eight chimp sanctuaries in the United States, and the opportunity to work with our closest living relatives was something I couldn't pass up. Chimps Inc. had just four individuals, and I was excited to meet them. I packed my bags and traveled three thousand miles across the country to start my new career.

As I made my way to Oregon, I envisioned what it might be like to work with chimpanzees. I had been working with various monkey species in Texas, so I had some experience with primates; I knew there would be great diversity in individual personalities and behavior. I was excited to discover the traits of these four individuals.

When I first arrived at the sanctuary, I caught sight of a large, boisterous male, Topo, in one of the aerial tunnels. He strutted his two-hundred-pound brawny body toward me. I was impressed. At first I thought this was a positive gesture, a greeting. Quite the contrary. Topo was anxious and upset because I had entered his home. I learned that he was the alpha male and his reactions expressed dominance and fulfilled his protective duties, but they were also a result of his unfortunate history with humans.

Topo had been shifted around from home to home, and through different sectors of the entertainment industry, during his formative years. He endured years of gawking humans, suffering humiliation while crowds of these other primates—*Homo sapiens*—laughed at him. I heard that one person crossed a barrier, defied the Dangerous Animal sign, and punched Topo. His natural reaction to this outsider's hostility was returned aggression, and the belligerent human was severely injured. Because most nonhumans who injure humans are therefore considered dangerous, Topo was scheduled to be euthanized.

But an animal trainer learned about Topo's situation, recognized an opportune moment to make a business deal, and bought the chimp. Topo was then incarcerated in a dark cell at this man's deplorable roadside zoo in upstate New York. He was robbed of any social or physical contact with other chimpanzees. His body only knew the touch of a cold concrete floor. His small, windowless cage deprived him of natural light, and the only fresh air he ever felt was when someone opened the door.

Once I understood Topo's display, I presented myself in a nonthreatening, almost submissive manner. I wanted him to know that I was not a danger to him or his chimpanzee family, and his swagger diminished. When I turned to look at him, he ever so slightly shifted his attention to his water bottle—as if exhausted by his grand display of aggression—and took a swig. "Success," I thought to myself; "I am making progress!" It was at this moment that Topo—his aim honed from years of practice—doused me with his water, squirting it from his puckered lips like a miniature fire hose. In that moment, I realized that gaining the trust of these cousins of mine—let alone their friendship—was not going to be easy.

This hazing—or initiation—went on for quite a few months. The chimps challenged me at every door of their enclosure and ignored me with every ounce of their being. They spit on me and showered me with the delicious smoothies that I had meticulously prepared for them. Every day was a new adventure in patience and fortitude. I just kept reiterating what the former caregiver had said to me: "In time they will trust, and that will be your reward."

Eventually their unenthusiastic behavior lessened and was replaced with more positive gestures. They used American Sign Language to invite me to chase them and engaged me in one-on-one grooming sessions. Each of these changes was momentous in my daily routine; I felt as if I had been invited into their family circle. My relationship with them grew stronger every day, particularly with Herbie.

Herbie and his companion, Kimie, arrived at Chimps Inc. in 1998. Both chimpanzees had been discovered in a small shack, made with chicken wire, at a private residence in Lebanon, Oregon. They were malnourished and lacked many basic, natural chimpanzee mannerisms. When Herbie first arrived at the sanctuary, he had to be kept separate from the existing social group because whenever he saw another chimpanzee (besides Kimie) he panicked and screamed. Even today he has little knowledge of the way to deal with his own aggression and is terrified when other chimps display natural belligerence. He lacks social skills—submissiveness and reconciliation behavior—that most chimps learn when they are very young. Herbie was separated from his biological mother at an early age, raised by humans, and shuffled through multiple homes during his formative years. Consequently, he did not learn how to get along with other chimps, and this makes it difficult for him to integrate with others of his kind, particularly males. This is a common problem for captive chimpanzees.

Herbie and I became very close during my second year at the sanctuary. He had both a morning and an evening ritual. Every morning he pleaded for my attention by banging on his door. When he saw me, his thick black hair stood upright on his body, much like a little kid with goose bumps on his arms from excitement. When I approached, I could smell Herbie's distinctly pungent body—he smelled almost like a freshly cut, sweet onion. He then pulled out one of his favorite ball caps and passed it to me through the bars of his enclosure. Naturally he wanted me to wear this cap. When I placed it on my head, he bobbed his head up and down and clapped his hands with delight.

Each night, before I turned off Herbie's lights, he called me over and waved his index finger at my feet, asking me to take my shoes and socks off so that he could groom my toes. When I granted his wish, he opened his mouth wide and breathed heavily—a chimpanzee expression of joy and enthusiasm. Herbie gives great pedicures, and I suspect these grooming sessions would never have ended, even if I had stayed all night. Unfortunately, there always came a time when I had to turn the lights out and say good night. I felt like a parent tucking a child in before bedtime, wishing him sweet dreams. When the lights went out, Herbie let out a gentle hoot as if he was saying, "Good night; see you tomorrow."

In 2005 I received a letter from a graduate program on the East Coast, re-
minding me that this was my final year to defer acceptance to its ecology program. I had postponed my admittance into the master's program because I
wanted to gain additional hands-on experience and determine exactly the nature
of my study. When the school offered me a full scholarship, I knew I had to accept, even though I would have to leave my best friends—my chimp family—in
just two short months.

My last day at the sanctuary—and the first day of a new chapter in my life—
finally came to pass. Although I knew, without a doubt, that this new beginning
would have a profound and lasting effect on my future, leaving was emotionally difficult. When I offered my final good-byes to the chimps, they wouldn't
even look at me. I called to each one over and over again, but no one wanted to
see me. They were familiar with the preparations humans go through when they
leave for a long time: the bags, the packed cars, the good-byes. Most often their
human-adopted friends never came back. I remembered their hesitation to accept me when I had first arrived. Was I betraying their trust?

My final call was to Herbie. I looked atop the aerial tunnels, under the blankets, and in the towers. He was nowhere to be found. Like the others, Herbie
knew that I was leaving. I had broken the circle of trust that I had worked so hard
to build. That is how I left my chimpanzee family—heartbroken—with unspo-
ken good-byes and tear-filled eyes.

Days later, I thought to myself, "What could those chimps possibly be
thinking? They must hate me." Then I received a phone call from Lesley Day, the
president of Chimps Inc. Her mind was thinking faster than she could speak, so I
had difficulty understanding her. Finally, she settled down and told me her good
news:

> Paula, I went into the Chimp House today with my hands filled with pa-
> perwork: pictures from your going-away party, bananas, and grapes. Her-
> bie kept calling to me and pointing to the stack of things I had brought. I
> assumed that he wanted the bananas, so I passed him one through the feed
> door. Though he didn't ignore the banana, that was clearly not what he
> wanted. He continued to point. So I thought, "Ah! He wants the grapes."
> So I passed Herbie the grapes. Again he accepted the grapes, but he set
> them aside and continued to point to the other items I had in my hands. I

finally threw up my arms and said, "I have no idea what you want!" I took everything I had brought and passed it to him. He immediately started shuffling through all the pictures from your going-away party. He found a picture that had just you in it and pulled it from the pile. He then carried the picture up onto his bench and stared at it for the rest of the day.

I had spent considerable time and energy convincing myself that the chimpanzees didn't understand how much I cared for them, that I had been there for a couple of years and then gone away—just another human. I had convinced myself that the bonds I perceived between me and the chimps were merely anthropomorphism, emotions that I had created to make myself feel as if I was connecting with them. But Lesley's story spoke volumes. I was wrong. Herbie was displaying what I was feeling—sadness, despair, longing, loss.

I learned later that Herbie carried my picture around for days. Whenever he moved to a different enclosure, he took my picture along with him. Herbie's undeniable connection offered a pivotal moment, a lesson in friendship and trust, and more importantly, a lesson in steadfast, unconditional love. He forced me to grasp deeply the fact that time is finite. Life is to be lived, and living is about life's simple enjoyments, whenever and however they manifest themselves.

I completed my master's degree in ten short months, then gratefully accepted the executive director's position at Chimps Inc.—permanently. I took this job not to advance my career but to rekindle the relationships I had built with my chimpanzee friends.

I once again traveled three thousand miles across the country to work with chimpanzees, but this time I was greeted by a large, boisterous male—Herbie. He gestured to my shoes with his index finger and hooted with excitement when I wiggled my eager toes in front of his ready hands, awaiting a long-delayed pedicure. Herbie reminded me that friends expect us to keep faith with one another.

17

¡Comejenes y Terremotos!
(Termites and Earthquakes!)

Keri Cairns

Headed for Ikamaperu

In 2003 I was asked to help at a Peruvian monkey-rescue center, Ikamaperu (http://www.ikamaperu.org/). The owners had to go away for a few weeks and needed an experienced woolly monkey caregiver to cover. I arranged for six weeks off work and headed for South America. Just before I lifted off from the United Kingdom, the Kilverstone Wildlife Charitable Trust granted three thousand pounds to Ikamaperu, which provided an unexpected challenge for me.

Since 1998 Carlos and Hélène Palomino have been looking after rescued animals at their home in Moyobamba in the San Martin region of northern Peru. One day an unidentifiable bald monkey was given into their care; he was badly malnourished and covered with engine oil. After weeks of intensive care, they were able to identify Taysu as a brown woolly monkey (*Lagothrix lagotricha poeppigii*). When I arrived, he was about four and a half, full of life, and was in the company of eight other rescued woollies. They shared an enclosure in Hélène and Carlos's back garden, from which they emerged during the day—accompanied by their keepers, Gardell and Amirio—to enjoy the freedom of the ravine at the bottom of the garden. Here they were able to forage for insects and leaves and enjoy a rich seasonal bounty of fruits.

Carlos and Hélène owned forty-four hectares of land along the banks of the Rio Mayo, known as Tarangue. They were working with the local Aguaruna Indian community to regenerate the land with various wild fruit trees. Thanks to the newly arrived, generous grant from Kilverstone, my job was to design and build suitable sleeping enclosures at Tarangue, ones that were large enough for the monkeys to live in during the rainy season. This seemed simple enough until I learned that these enclosures also had to be earthquake and termite proof!

Amazonian woolly monkeys live in colonies ranging from a few individuals to sixty monkeys; there is even an unconfirmed report of a group of one hundred. They are a beautiful species with luscious fur ranging from gray to brown and black. They have a powerful prehensile tail that enables them to hang upside-down, freeing their hands for picking fruits too large for other primates.

Woollies are some of the heaviest South American nonhuman primates. Unfortunately, this means that their meat is highly prized, and they have become a prime target for hunters. Adding to their problems is a lucrative illegal pet trade; a baby woolly smuggled into North America can fetch as much as twenty thousand dollars. However, as with most primate species, their gravest threat comes from the destruction and fragmentation of their habitat.

Vast swathes of the Amazon are destroyed every year for a variety of reasons. Loggers make a lot of money selling hardwoods such as mahogany; they also use a lot of the larger trees to produce plywood and medium-density fiberboard to supply our Western appetite for do-it-yourself projects. The forest is also cleared for agricultural purposes such as grazing beef cattle or growing soya to feed those cattle for market. I became a vegetarian twenty-five years ago after reading about the effect that this production of cheap beef, mainly for Western fast-food outlets, was having on the Amazon rain forest. Wild woolly monkeys only live in primary, untouched forests, rather than secondary ones that have been cut and replanted.

Woollies play a vital role in the Amazon by dispersing seeds. Many of the fruits or pods that they eat contain small seeds, which pass through their intestines and are deposited on the forest floor in a ready-made compost package. Dung beetles then bury this seed-laden compost; a seed that has passed through a monkey germinates faster than one that simply falls to the ground. A study in Colombia in 2000 states, "Given a population density of 30 individuals/km^2 [per square kilometer], the woolly monkeys in the study area disperse more than 25,000 seeds/km^2/day. These seeds belong to 112 different plant species" (Stevenson 2000, 275). This makes woolly monkeys and other large nonhuman primates essential to the diversity of a healthy forest. A study that compared a protected forest to one where hunting was permitted showed a marked reduction in the diversity of tree species in the unprotected forest. Therefore, safeguarding primates safeguards forests, and safeguarding forests safeguards primates.

Although woolly monkeys are officially protected, their bodies provide a large proportion of the bushmeat available in local markets in Peru. One of Hélène and Carlos's friends traveled to the market town of Yurimaguas where several large rivers converge, generating considerable trade. She returned with photos of smoked woolly monkeys—entire colonies—openly for sale. The police were present but did not take any action against this illegal activity. Even more shocking, these markets were selling live baby monkeys alongside the tables of smoked meat.

I flew into Tarapoto from Lima over the snow-capped Andes with their seemingly random roads and scattered little villages. The mountains dropped off abruptly when we reached the Amazon rain forest. Just as abruptly, the forest disappeared and was replaced with unending miles of rice fields and palm plantations. I could see the fires smoldering as locals turned rain forests into grazing land to provide meat for Western markets. Amid the fields, we descended onto a large strip of tarmac.

I had been expecting a humid atmosphere, but upon stepping off the plane, I felt only dry heat and dust. I was met by a friend of Hélène and Carlos's, who showed me where to exchange money and catch my ride to Moyobamba.

From Tarapoto to Moyobamba, I traveled for two hours by autocolectivo (a shared taxi) along a large highway. I was startled by the lack of forest, which had been replaced by plantations of maize and rice and occasional fruit trees. There were mounds of charcoal across the land and a constant smell of smoke. The roads were lined with people carrying piles of wood and cane. As the autocolectivo climbed into the mountains, the air cooled, and the hillsides showed signs of serious erosion. Only a few patches of trees remained, clinging to steep cliffs—inaccessible to loggers.

When I arrived at Moyobamba, Hélène and Carlos greeted me, but I had to wait until the next morning to meet the monkeys. It was amazing to see them out and about, taking full advantage of their lovely ravine. For the next week, I enjoyed the company of Hélène and Carlos and learned the ropes of the sanctuary. Then they were off, and I was left to explore and experience the delights of the woolly monkeys.

Apu–El Jefe (The Boss)

An adult male woolly monkey can be extremely dangerous. In the wild, they rarely come into contact with humans, except through hunting. When dealing with woollies in captivity, caregivers must never enter an enclosure containing an adult male. They are very territorial after they reach maturity—between six and seven years of age. At this time, they also develop large canine teeth and a large muscle around their head, which attaches to a sagittal crest, giving them incredibly powerful jaws.

One day a number of years ago, Carlos and Hélène received a phone call from the local police station. The officers wondered if they could come to collect two woolly monkeys. One of them, Apu, had already attacked three police officers before Carlos arrived. So it was that Apu and two-year-old Shayu came to live at Ikamaperu.

It was difficult to interpret Apu's history; he was a bit of a mystery. Though fully grown, he was slightly smaller than other male woollies I had worked with; I think he weighed around eight kilograms (17½ pounds), whereas there are reports of wild woollies reaching fourteen kilograms (31 pounds). He had the long chest fur of an adult male, but he did not possess the male woolly scent; males I have known all produce a strong musky smell from a gland on their chest that they use to mark their territory. It seems very unlikely that he would have survived to reach maturity as a pet, but it appears equally unlikely that he was captured alive as an adult since full-grown male woollies are fierce. Apu had several scars on his head. These could have come from either another male in the wild or a savage human. Also his left-top canine tooth was missing; again this could have resulted either from age or human violence. He was accustomed to people so long as they were polite and respected his personal space, and he was incredibly protective toward Shayu. How had these two come together? Where had they come from, and at what point had they come into contact with humans—and one another?

Many pet monkeys are released back into the forest when the novelty of having a furry little primate wears off or they start to become aggressive as adolescents. It is possible that Apu's mother and her colony had been killed when humans captured him at a very young age. In that case, he might have been kept as a pet for some years and then released. While released primate pets rarely survive—having no natural survival training—he might have joined a small group of woollies who looked after him. This might have been where he met Shayu. And if he had been treated relatively well by humans, he might have continued to associate people with food. If so—due to the destruction of habitat or perhaps bushmeat hunters who killed the rest of his colony—Apu might possibly have returned to humans in search of food, eventually being arrested.

At first I was quite nervous around Apu. Before he was allowed out of his enclosure, we tied a belt with several yards of rope around him. This was for his own safety since he might wander off and get into trouble. He tolerated the rope, patiently untangling himself when he climbed trees. The monkeys' keeper, Gardell, had built several bamboo platforms at the bottom of the garden and down into the ravine. This allowed caregivers like me to climb up into the trees and hang out with the monkeys while they foraged and socialized.

When I had previously worked at the Monkey Sanctuary in Cornwall, some of the females and youngsters had been allowed into the gardens. I was trained how to behave around woollies, and I soon learned to read their body language, some of which is very subtle. It is particularly important to avoid direct eye contact with woollies, especially adult males, because a direct stare is interpreted as a threat. As a result of my previous training, I behaved politely, and the monkeys at Ikamaperu began to trust me very quickly. I followed their social mores and meticulously avoided any threatening behavior. I even adjusted to Apu quickly, even though I had seen him rip off a door between two sections of his enclosure.

Apu showed a lot of natural woolly monkey behavior, likely the result of spending considerable time with a wild colony. In their natural environment, adult males help to administer discipline and enforce acceptable social behavior. If Apu saw the younger monkeys misbehaving, he gave them a hard stare or even a good shake, depending on just how naughty they were. In the wild, adult males also help to look after youngsters, even sometimes giving them piggyback rides.

Apu and I developed an understanding, and toward the end of my stay, we shared several fascinating interactions. There were large coconut palms in the garden where the woollies enjoyed outside time. Every few days Gardell shuffled up one of the trees and tossed down some coconuts. We chopped off the outer husks with a machete, then broke the coconuts into smaller bits for the monkeys. This took me a while to get the hang of; I had to be very careful with the machete since the monkeys clambered over me in search of coconut treats. Apu always waited patiently, knowing that I would give him the first bite. This was important because adult males always eat first.

Woolly monkeys spend much time play-fighting; youngsters' energetic and boisterous games are accompanied by a chuckling sound. The aim of the game is to touch the back of your opponent's neck and ideally administer a playful bite. The same goes for the tail, which has a very sensitive palm of skin over the last six inches. Adult males, however, play a very different game. They expose their teeth with a wide-open mouth and a gentle shaking of the head, accompanied by a quiet chuckling sound. Adult males rarely play, but when they do, their actions are very gentle so as not to provoke confrontation. They are incredibly strong and could easily do each other serious damage—an adult male hanging by his tail could lift a fully grown man off his feet!

One day, after he had enjoyed his piece of coconut, Apu remained seated by my side. He seemed relaxed, and I decided to take a bit of a gamble. I looked at Apu and put my head back, opened my mouth, and gently chuckled. He watched intently. I gingerly grabbed and gently nibbled the end of his tail. Much to my amazement, Apu opened his mouth and chuckled back! Then he placed his hand on the back of my neck, but faced with those giant canine teeth, I decided not to push my luck. I changed my behavior to initiate grooming. Apu stretched out and pointed to places where he wished to be groomed. Primates strengthen bonds through grooming, and so I felt incredibly privileged.

Inside the monkeys' enclosure was a hammock where I sometimes stretched out so that the youngsters could come to play on me, or just curl up and enjoy a groom. One day, shortly after that first, privileged grooming session, Apu saw me grooming young woollies on the hammock and strolled straight toward me. I kept on grooming and was surprised when Apu lay on top of me in the hammock! This would have felt very unsafe with any other adult male woolly that I knew—particularly because I was enclosed in his territory—but I recognized that we had reached a new understanding. I had won his trust, even though he was wary of humans.

Building a Safe Haven

Eventually I turned my focus to my building project and started to spend more time at the Tarangue reserve. Although Tarangue was mainly secondary forest, there was a section of primary forest where wild monkeys lived. It was near there that Amirio and I cleared an area for the new enclosure. Once we had prepared the site, I asked Amirio if we might go in search of the wild monkeys.

Though Amirio told me that I clumped through the forest like an elephant, I was nonetheless rewarded with some distant views of the monkeys. At first I thought that they were tamarins, but when I later looked at my photos, I realized that they were Andean titi monkeys (*Callicebus oenanthe*). These monkeys are listed as threatened, but several researchers recommend that they be reclassified as critically endangered; they exist only in the Rio Mayo valley. Tarangue is one of their last strongholds, and it is obvious that their main threat is human fragmentation of their habitat. I walked into a slashed and burnt area just five minutes from Tarangue—a blackened mass of smoldering trees and scorched earth. Most of the area suffers from slash and burn—people destroying rain forest to create grazing and cropland.

Hélène and Carlos's friend, Rolando, helped me select materials for the woolly monkey enclosure at Tarangue. First, he took me around Moyobamba, pointing out different local constructions. The preferred building method included lots of concrete, which seemed odd when so many concrete buildings had been destroyed by two powerful earthquakes that hit Moyobamba in 1990 and 1991. In contrast, Hélène and Carlos's house had withstood the earthquakes and was made of traditional materials: a wooden framework with walls of cane.

One day, when Rolando took me to a government building where he had a meeting to attend, I had an important insight. The government building was one of the largest in Moyobamba and had previously been residential, but people now refused to live in any building so likely to collapse in an earthquake! While I waited in the canteen, I realized that termites only infest damp wood, and wood is generally damp where it comes into contact with the ground. I decided to include a simple gap between the wood and the concrete support in hopes of solving the termite problem. This would also allow the whole enclosure to be wooden, and it would flex in the event of an earthquake and stay standing long after the government building had collapsed in a heap of cement blocks and dust.

It was then that I realized that our biggest problem was not the threat of an earthquake, or even the fearsome teeth of Apu, but the fact that the grant money we needed to build the enclosure was not yet in hand. I was running out of time. Luckily—when I had only two weeks left in Peru—the funds arrived, and

Gardell and I headed straight to Moyobamba to order materials. We also hired a boat to carry the materials to our chosen site in Tarangue, and I met with an engineering professor, Romolo, to ensure that the project would be properly finished after I returned to the U.K.

Romolo explained the mechanics of earthquakes: the problem is that the earth sometimes sumps under one corner of a structure, which pulls the whole building down, even if the materials are flexible. He suggested a completely concrete foundation strengthened with metal bars, which did not appeal to me. This would drive the costs up and also require transporting a lot more materials down the river.

I traveled to Tarangue with Romolo the next day and was relieved when he noted that I had picked the perfect location for our enclosures—far enough from the river so the land would not give way. I was also pleased when he agreed to carry on my original plan.

We started work at 6:00 a.m. the next day, and everything was loaded onto the boat by 3:00 p.m. Any more weight, and we would have been too heavy to float. We acquired a massive audience; everyone wanted to know what the crazy gringo was up to. We headed downriver, and everything was unloaded by 5:30 p.m., allowing just half an hour of daylight for Gardell to return the boat. We had planned that Amirio would come along later to spend the night, but in the meantime, I was alone. I went to the kitchen and put on some rice. When I looked up, I saw lights where we had unloaded the materials. I grabbed my headlamp and strolled toward the wood, making lots of noise. Whoever was there left as I approached, and I could hear a boat bumping on the riverbank just around the corner and whispered voices.

I suppose I should have been frightened, but I had only one thought: no one was going to steal the materials that offered such hope for the lively monkeys I had come to know at Ikamaperu. I turned my lamp off and sat down by the river. The mosquitoes swarmed and whined all around. After an hour, I heard the prospective thieves start their outboard motor and head downriver. I gathered up firewood, then lit a fire on the riverbank, thinking this would deter any other thieves. As I set off back to the kitchen, I heard someone approaching on the path. Thankfully it was Amirio! He had arrived at the kitchen, discovered a pot of burned rice, and wondered whatever had happened to me. When I explained, he agreed that the fire was a good idea and said that I was, indeed, a crazy gringo for approaching thieves.

We awoke early to begin our work and were happy to see that the materials were all accounted for. Manuel, a local carpenter, and a few others came to help. I was amazed at their resourcefulness. Manuel dug a well to provide clean water for the concrete. The others collected rocks and extracted sand from the riverbank to mix with our cement to create the necessary concrete. I got to work making hatches that would separate enclosures if one of the woollies became ill or for occasions when new monkeys were introduced to the group. Two of the hatches were placed high enough so that the monkeys could leave the enclosures and go straight into the trees of Tarangue. The hatches were operated remotely with pulleys.

I had special metal plates made—platinas—that created a gap of three centimeters (one inch) between the cement and the wood to prevent it from becoming damp, and I painted pitch on the bottom of the posts for further protection. I kept the others very amused with my marginal Spanish, bumbling through their language in innumerable ways, perhaps most famously by mispronouncing platinas as platanos—"bananas."

When I left, just over a week later, the Tarangue enclosures were progressing nicely. Unfortunately, Manuel didn't quite understand my plans for the hatches, so my vision of two enclosures ended up as one large enclosure, and only the remotely operated hatches were ever used. (I heard that when Manuel was later constructing an accommodation for Gardell and Amirio, he built the walls around him and forgot to put in a door, leaving no way out!) Nonetheless, the monkeys were moved to Tarangue about six months later. I have been told that they love their new home.

In 2005 Moyobamba was hit by an earthquake measuring 7.5 on the Richter scale, larger than either the 1990 or 1991 quakes, and the Tarangue enclosure survived!

Looking Back on Project Tarangue

There are now forty woolly monkeys living in the Tarangue sanctuary, along with a few needy spider monkeys and capuchins, and two yellow-tailed woolly monkeys—a critically endangered species found only in the cloud forest north of Moyobamba.

For decades people had assumed that the yellow-tailed woollies were extinct until they were rediscovered in 1974. In 1999 estimates placed the number of yellow-tailed woollies at less than 250, and they were placed on the list of the twenty-five most-endangered primates in the world. Thankfully Neotropical Primate Conservation, a newly formed organization (http://www.neoprimate. org/), is working closely with Ikamaperu to help save yellow-tailed woollies from extinction.

Hélène and Carlos have recently acquired lands that border a national park, home to wild populations of woolly monkeys; some of the rescued woollies have been relocated to this area with the hope that one day some of them may be reintroduced to the wild.

Sanctuaries such as Ikamaperu are essential in primate-habitat countries because they offer safe haven to these unfortunate refugees, whose lives have been disrupted by human interests and indifference. There are many sanctuaries around the world that help primates, but they cannot do this essential work without help—your help. Most sanctuaries run on a shoestring budget, and financial assistance is always needed. So why not make a donation to an individual sanctuary, or to IPPL (International Primate Protection League), which provides much-needed financial support and advice. Many sanctuaries run volunteer programs that allow visitors to help out. There is a fee for this unique life-changing opportunity, and the funds go directly to the sanctuary.

I remember my time in Peru with great fondness. I am blessed to have known Apu and spent such days among nonhuman primates. I am proud that I was able to help a sanctuary that is working hard to save species like the yellow-tailed woollies and individuals like Apu.

Sadly, several years after moving to Tarangue, Apu succumbed to a chest infection, but I am comforted to know that his final years were good ones. I hope to return to Peru someday to see just how much Ikamaperu's fledgling monkey colony has grown; I know they will thrive in such good hands. I feel fortunate to have been able to help out at sanctuaries, including Ikamaperu, and to have met amazing characters like Apu whom I regard as some of my closest friends.

18

—

Singe

Helen Thirlway

My relationship with Singe did not exactly get off to a good start. I had volunteered to run the Monkey Sanctuary in Ireland for a few months, along with three other caregivers, so that the owner could enjoy a much-needed sabbatical. Although we two women were allocated the task of caring for Singe, we were told that this little nonhuman primate did not like other females. This was only too true. Whereas a male caregiver entering her quarters elicited snorts of happiness and shuffling forward for a groom, Singe barely tolerated Marianne and me, and a quick movement or too much eye contact resulted in a beady-eyed stare, followed by a threatening leap in our direction. Friends we were not.

Singe was a house monkey—she had lived in a human home in Europe for some twenty years. I do not mean that she should have been in a house—in fact, quite the opposite—but, sadly, years as a human's pet meant that rehabilitation to a more natural, monkeylike existence simply was not possible. Years of sitting on a sofa eating junk food had made her obese and arthritic.

Most of our sanctuary residents were capuchins, a small and hardy species hailing from South America. The majority were tufted capuchins (*Cebus apella*), little brown monkeys recognized by their distinctive head with a black or dark brown cap and dark sideburns. It is this cap—which was thought to resemble the cowl worn by the Capuchin monks, an Italian order founded in the 1500s—that gave them their name. Some were ex-pets, but the majority of the capuchins had come from a behavioral research laboratory. A few had been caught in the wild, and they taught the others how to behave in the more natural environment of the sanctuary. Despite many of them having been confined, since birth, to bare cages in a room without windows, they now lived fairly contentedly on human-made soft-substrate islands, where lush foliage provided plenty of foraging. There was very little need for caregivers to provide enrichment for these capuchins because watching and shouting at other monkeys across the water, ganging up on swans and ducks, and climbing and foraging in the trees kept these busy little residents constantly entertained.

But Singe was a De Brazza's monkey (*Cercopithecus neglectus*), native to central Africa and named after the French explorer Pierre Savorgnan de Brazza. There are no sanctuaries in Europe for this species—at least none that we could locate—and no zoo would take her. So here she was at a respectable sanctuary, but alone, with no companions of her own kind.

When she was moved to one of the moated islands at the Monkey Sanctuary, it must have seemed a strange environment to her with no suitable companionship. The stress caused Singe to hurt herself to the degree that medication was required to heal her self-inflicted wounds. Attempts to house her with primates of different species did not work, either. She barely knew how to be a De Brazza's monkey, let alone interact with individuals from another species who had a different language and distinct social etiquette. Eventually, after many painful efforts, we realized that rehabilitation was not possible, and she was moved to an indoor setting to live out her days in what had become, for Singe, a normal—and thus less stressful—environment. A room adjoining our kitchen became her bedroom with a basket for her bed and a heat lamp for warmth—a vital element for a primate from Africa displaced to Europe.

Every morning I reported for duty in the kitchen to prepare Singe's breakfast. I called to her to let her know I was there. Usually there was no response, much to my frustration; I knew that the men's voices always prompted a snort of delight, followed by Singe's tubby, diminutive body poking through the doorway. Singe had the distinctive appearance of a De Brazza's monkey: the speckled olive gray coat, white lips and beard, and a bright orange crescent-shaped tuft above her head—although on Singe, some of this had been plucked out during periods of stress, leaving a bald patch. But her resemblance to a normal, wild De Brazza's monkey ended there. Singe was fat; she didn't walk; she waddled. A diet was clearly in order. I experimented to find out which substitute foods she enjoyed most, and to my surprise—despite her previous junk-food diet—she liked nothing better than green salad leaves with fresh wild herbs or nutritious seeds.

She still had a sweet tooth, however. One day she spotted a big multipack of Tic Tacs in the kitchen. When Marianne and I were distracted, she deftly climbed up and grabbed the whole pack and ran off to her room. It was amazing how fast and nimble she could be when she really wanted something, especially considering how slow and frail she appeared the rest of the time. We followed her quickly but with trepidation because she was quite formidable when she was determined about something! Bravely, Marianne did manage to pry the multipack out of her greedy little hands, but only after she had already managed to break open one of the packets and wolf down the entire contents. At times like these, Singe was the epitome of the cheeky monkey. It was less amusing when we discovered that mints eaten in large quantities have the same effect on monkeys that they do on humans. I will not go into detail, but suffice it to say that cleaning up after a monkey that has eaten a whole pack of mints is not a pleasant experience.

Having witnessed Singe's ability to forage in a domestic kitchen, I emptied the rest of the Tic Tac boxes and filled them with nuts, seeds, and dried fruits, then hid them around the kitchen in places she could find with a bit of work. It took her weeks to locate all of them, and when she found one, the box kept her happily working to extract the raisins and nuts for quite some time—they were not as easy to dislodge as the Tic Tacs.

In spite of my efforts to provide an interesting diet and create new forms of enrichment and regardless of the hours I spent sitting with and talking to Singe, I made little headway in the friends department. Or so I felt. She did eventually ask me to groom her. Well, demand is probably closer to what actually happened: she made the request by sitting next to me and pulling my hand toward her. I was particularly taken aback when—while grooming Singe—I became distracted talking to one of the other caregivers and paused. Within minutes Singe was grabbing at my hand and placing it on her back to make it quite clear that her grooming session was not over. After that I got used to feeling her familiar, determined tug whenever I lost interest or got distracted while grooming her. It was obvious who was on top in this hierarchy.

Her aggression made me wary, which didn't help our relationship since I was jumpy and on edge, worried that she might carry through with one of her threats and bite me. De Brazza's are a species of guenons, and these monkeys make many varied facial expressions when they are excited or angry, including exposing their teeth, bobbing their heads, and staring or closing their eyelids. In fact, the different grimaces may explain how they acquired the name guenon, which was in usage in seventeenth-century France to refer to a "very ugly woman." Singe never did actually try to bite me, but her beady-eyed stare and open-mouthed threat served as fair warnings.

Finally one day after about two months, rather than demand a groom, Singe came to me when I was kneeling on the floor and started lip-smacking, which I had never seen her do. Her lip-smacking made me curious about her background. She must, presumably, have learned this behavior from her mother. So how old was she when they were separated? Was she born in captivity or in the wild?

Having approached with friendly overtures, Singe stared at my bare arm, studied it carefully, then pulled gently at the hairs on my arm and put imaginary skin (or mites!) into her mouth. She only groomed my arm for a few seconds on that wonderful day of change, but after that, her friendly grooming sessions became progressively longer.

Before my experience with Singe, all my work at sanctuaries had been hands off—as it should be in all but the most exceptional cases—to allow monkeys to interact naturally with their own kind. Sometimes a fully trained and experienced caregiver might share a bit of grooming through enclosure wires, but only in exceptional circumstances. I had occasionally cleaned an enclosure that held both female and baby monkeys, at which time I interacted a little with the mothers and their young. Nevertheless, I had been trained to see nonhuman primates as wild, dangerous, and unpredictable—which they most certainly are. Consequently, I was awed by Singe. Eventually she started clambering onto my lap to groom my neck and even tried to remove items of clothing to "clean" my skin more thoroughly.

As my three-month stay at the sanctuary came to a close, Marianne and I dreaded leaving Singe. It was now we whom she rushed to greet when she heard our footsteps and voices. It was we whom she groomed most intimately, and it felt all the more special because we had worked so hard to develop this trusting bond.

I had never been so challenged by a monkey as I was by Singe. There were times when I felt—dare I say it—that I hated her. It was exasperating to spend so much time and energy trying to win this primate's trust only to be repeatedly threatened. Now she was our friend, but in the world where she had been placed—the human world—her relations with caregivers were neither equal nor dependable. Though we tried to provide her with as many of her needs as we could, she remained a stranger and a captive, and ultimately we had to return to our previous jobs and leave Singe behind. Luckily her snorts of pleasure on seeing the sanctuary owner, Willie, when he returned, reassured us that we would not be too sorely missed.

I believe that Singe could have been rehabilitated to live with other De Brazza's monkeys if there had been an opportunity for her to do so. She was only humanized to the extent that she had become used to living in a house and eating unhealthy food. But watching her forage, lip-smack, and groom, I could see that there was no mistaking the true monkey within—she certainly suited her name, which simply means "monkey" in French.

My strongest and happiest memory of Singe is the moment when she first climbed into my lap and fastidiously groomed my neck. It is hard to describe the complex range of emotions I experienced, sitting in the kitchen with this little, lonely and isolated, damaged and neglected monkey treating me like friends and family. I felt excitement, nervousness, and love, but there was an all-pervading sense of sadness—sorrow and loss for the Singe who could have been—the one who would have spent her days grooming family members, climbing and foraging in the trees, playing, hunting, and thriving wild and free. This was the life that had been stolen when Singe (and/or her mother) was taken from the wilds of Africa and brought to Europe to be bred and sold like a commodity so that people might enjoy the thrill of having an exotic pet. Though I treasure my time with Singe, it was certainly not worth the price that she paid.

The Primate Pet Trade

Currently it is legal to buy, sell, and keep monkeys as pets in Ireland and the United Kingdom, the majority of European Union member states, and much of North America. Most primate pet owners are not deliberately cruel—just misguided—and many realize too late that this practice is not appropriate and should certainly not be legal.

Primates are wild animals. Many of the monkeys kept as pets would naturally range across huge areas of forest in groups of up to forty individuals. In the wild, most of their day is spent foraging for food and socializing in the company of their own kind with an occasional need to hide or run from a predator. They have evolved to be extremely intelligent. They have adapted to their environment and the complexity of living as part of a large social system. Capuchins, one of the most popular pet species among primates, are even known to fashion and use tools, and they are the only monkey—rather than ape—to have demonstrated this ability.

Taking such highly evolved creatures out of their natural habitat deprives them of their mental, social, and physical needs. Monkeys kept as pets are often malnourished due to inadequate diet and lack of sunlight. Monkey pet owners unwittingly deprive these intelligent primates of the psychological and environmental stimulation that is so important to their welfare. Consequently, most pet monkeys display stereotypical behavior, such as pacing, rocking, overgrooming, and self-mutilation. Human children who have been similarly mentally and socially deprived during their formative years practice similar behaviors.

People who buy baby monkeys are separating them from their mother at an extremely young age, causing trauma for both the mother and her offspring. Many people are attracted to baby monkeys because they are young, gentle, and vulnerable. However, once monkeys reach adolescence, they naturally start to test their position in the group hierarchy and behave aggressively. It is often at this stage that monkeys must be sent to a sanctuary. Unfortunately, most sanctuaries are full, so the monkeys have nowhere to turn for refuge. Furthermore, only a few primate species are represented in a single sanctuary, so even if there is room, many individuals must live alone the rest of their lives.

Monkeys who are sold as pets in their countries of origin have usually been caught in the wild. This means that their mothers have been shot so their babies could be captured. This trade in exotic pets clearly harms individual primates, family units, troops, species as a whole, and ecosystems. The pet trade has devastated whole populations in certain areas and pushed primates all over the world to the brink of extinction.

What sort of message are we sending people in these countries when we keep monkeys as pets in the West? These intelligent, social, and extraordinarily athletic individuals are deprived, lonely, and displaced in human homes and then, if possible, foisted off onto sanctuaries that are already filled to capacity. It is time for all nations to pass legislation banning people from buying and selling our primate cousins as if they were pets.

What You Can Do

Please never buy a monkey, and try to dissuade others from buying any primate. Instead, think about adopting a monkey or ape at a reputable sanctuary, where your financial support can help a rescued primate thrive. When you adopt a primate at a sanctuary, you don't take him or her home; you simply underwrite the cost of caring for that individual. To help protect and conserve primates all over the world, you can also become a member of the International Primate Protection League (IPPL, http://www.ippl.org).

If you are keen to work with monkeys or other primates, there are sanctuaries all over the world that eagerly accept volunteers and interns. Please see the appendix to choose from the many worthy organizations that await your assistance.

19

—

Soiled Hands

Sangamithra Iyer

Hands...it was their hands that reminded me of what had happened there. It was 2005, and I was almost twenty-eight years old, sitting in a restaurant in Rwanda's capital, Kigali, with my friend Rachel. We had been traveling around the country for a month. On our last night, I found myself staring across the room at two deaf men using their hands to talk to one another.

I love watching hands tell stories. I studied American Sign Language (ASL) in college and learned the way fingers set up scenes and describe characters. The way the mouth and eyebrows corroborate what the fingers say. This was the language I learned to meet Washoe, the first chimpanzee to acquire ASL. Toward that end, I enrolled in a summer program at the Chimpanzee and Human Communication Institute in Ellensburg, Washington, before my senior year of college. Washoe greeted me as she did many new visitors: she smacked the thumbs of her two fists together, signing that she wanted to see my shoes. I looked at her hands and then looked at mine.

Rwanda is home to chimpanzees, thirteen species of monkeys, and the famous mountain gorillas in the mist. To see them in their natural habitat—on their terms—was partly why I was there. I had spent the previous several years seeking opportunities to visit and volunteer at primate sanctuaries as an escape from my steady environmental-engineering day jobs.

I had my dream job at the time of this trip: working at an environmental, social-justice, and animal-advocacy magazine called Satya, where I used my vacation to travel to Rwanda in pursuit of stories.

It had been seven years since I had first met Washoe, and I was still chasing apes. "Why do you care so much about them?" some people asked.

"I just do," I replied, not really knowing how I couldn't. Maybe it was because Tatu, another signing chimpanzee in Washoe's surrogate family, covered her eyes with her hands and signed "peek-a-boo." Or because Moja crossed her arms over her chest and signed "hug/love." Maybe it was because I played chase with Kiki Jackson at a sanctuary in Cameroon, and when we got tired, he presented his arm for grooming, and I pretended to find and eat his imaginary bugs. Or because I liked watching Nama's fingers lace and unlace the shoestrings of my hiking boots. Maybe it was because I made them lemongrass tea when they were sick and banana-leaf burritos when they weren't. Or perhaps it was because I held and bottle-fed the babies Emma, Niete, and Gwen in my arms. And right before equatorial sundown, when I bathed with just a few precious drops of sun-warmed rainwater, I could still feel where their tiny hands and opposable toes had latched onto my body.

Perhaps it wasn't so much these moments I spent with them, but rather their stories that touched me. They had been rescued from research, entertainment, and the bushmeat trade. Kiki Jackson was found emaciated in a concrete cell at a hotel in a coastal town in Cameroon with the body of his female companion, who had starved to death. He was a full-grown male chimpanzee, so we were separated by bars. I knew he was strong and could easily injure me unintentionally. I kept a safe distance and respected his space. But one day Kiki offered me his hand through the bars, and I held it in mine.

Little orphaned Gwen would grasp my shirt and refuse to let go. It was an embrace I felt long after her fingers relaxed, long after I was gone, long after she was gone. Seven years after my time in Cameroon, I received an e-mail informing me that she had been found dead at the sanctuary. The other chimpanzees were in shock. They crowded around making sounds of fear. One chimp pulled her body several feet in an attempt to wake her. Nama, the matriarch, started to groom her.

When I cared for Gwen, it was the first time I felt like a mother: this little life was completely dependent, and her human caregiver was the only mother she had. But the fear with which she clung to me reminded me that I was not the mother she was meant to have. Her mother had been murdered for meat when her little baby was less than a year old. Gwen didn't want to lose anyone else.

Meeting these nonhuman primates was a painful reminder of the horrors animals endure at the hands of humanity but also a chance to see them start new lives. I wanted to know that new lives were possible. I quickly learned that this was harder for some than for others. I was still haunted by the rocking of a chimpanzee—also named Rachel—that Rachel and I had met the year before in a sanctuary in Montreal. She spun her head in figure eights. Born at the Institute for Primate Studies in Oklahoma, she had been sold as a pet to a couple in Florida, only to be returned to a laboratory when she was just three and a half years old. Used in hepatitis research at New York University's Laboratory for Experimental Medicine and Surgery in Primates, Rachel lived alone in a cage for eleven years, was chemically knocked down with a dart gun 147 times for experimentation, and suffered thirty-nine punch liver biopsies. Not surprisingly, she had anxiety attacks and inflicted wounds on herself that caused rashes and sores on her neck. Though her laboratory days were long over, she still spun her head in figure eights.

Part of the reason I had come to Rwanda was to see free-living apes who had never known the loss associated with captivity or trauma inflicted by humans. But I was interested in something else, too. Eleven years prior, on the same soil where we stood, roughly a million people had been slaughtered in the course of one hundred days.

I had already come and gone from a chimpanzee sanctuary in Cameroon, where I had volunteered in 2002, when I started reading books about human violence. "Jeez, Sangu, don't you read any fun books?" my friends asked. I was, at that time, the definite buzz-kill at parties. I couldn't really explain my interest in the worst side of human nature—I just felt as if I had an obligation to know. On my daily commute on Metro North to my engineering job—crunching numbers in a cubicle in a corporate office park—I read about Rwanda's genocide in Philip Gourevitch's book, *We Wish to Inform you That Tomorrow We Will Be Killed with Our Families*. But I didn't know what to do with what I was learning.

As an engineer, I was interested in catastrophes—and in preventing them. I studied soil behavior because I wanted to understand the way the earth responded to human pressures. Trying to heal an already-damaged world was a challenge. A year before my trip to Rwanda, I was working on a Superfund project, the federal government's program to clean up hazardous-waste sites. But calling this project a cleanup wasn't accurate. Radioactive materials aren't something you can clean. They decay exponentially. We measure decay time in half-lives. For thorium it is on the order of billions of years. We couldn't clean it up, so the plan was to isolate the damage and relocate the contaminated soil—place it in one of two facilities in Utah that have been created to store low-level radioactive waste.

Rwanda had eight million people—the population of New York City. I tried to imagine what it would be like to have my hometown of New York decimated, and how life might go on. What is the half-life for the traumas of genocide? Unlike my Superfund site, the damage couldn't be isolated and removed. There was no Utah facility designed to store the toxic results of human violence. Perhaps I had come to Rwanda for the same reasons I've been chasing apes: to be reminded of what humans are capable of and see what happens afterward.

Shortly after we arrived in Rwanda, we met Pacifique at a church in Ntarama. The first thing I noticed about this slender man was the scar on top of his head, the place where a machete had once struck. I had seen a similar scar on Niete, an orphan chimpanzee I had cared for in Cameroon. It is hard to imagine hands responsible for such wounds.

During the genocide, people gathered at this church for safety, but they found no haven among these wooden pews or under this vaulted ceiling. Pacifique pointed to his head and simply said, "They slashed."

After the genocide, Rwanda's president, Paul Kagame, eliminated the use of ethnic identities. When people talked about the genocide, they couldn't say Hutu or Tutsi, but instead called them "they that killed" and "those that were hunted." Pacifique was left for dead. He survived, but he lost the rest of his family—in this church—that is now a memorial. This church in Ntarama is nothing like the memorials of sculpture and stone that I have seen in the United States. It is a memorial of wood and bone.

When we entered the church, we walked along the aisles and on top of the wooden benches, noting fragments of the five thousand people who had been slain here in 1994. The personal items of the deceased were also scattered about the floor: rosaries, little girls' shoes, scarves. The bones were sorted and stacked. Piles of femurs filled one corner. Skulls with vacant eye sockets gazed at us from nearby shelves. Skulls—too many to see—were also heaped in a bag against the wall. Little baby skulls were slashed in half. Who does this to a child? Little baby skulls slashed into multiple pieces. Who does this repeatedly to a child?

Rachel and I held each other and wept. The Gayatri Mantra, the only Hindu prayer I know, played on autoloop in my head; my instinct was to pray, but my prayers—like theirs had been—were inadequate.

The room adjacent to the church was a school. A child's notebook lay open on a small desk. Pages holding the cursive of little hands writing history flapped in the wind.

A few days after we left Ntarama church, we entered Virunga National Park, an area surrounded by three volcanoes: Sabyinyo, Karisimbi, and Bisoke. I knew these volcanoes from books, but in real life, they resembled the mountain gorillas who call them home. We had come to see the mountain gorillas in the gorilla mountains.

Straddling the borders of Rwanda, Uganda, and the Democratic Republic of the Congo, roughly seven hundred mountain gorillas are left in the world—each one is a magnificent and vulnerable individual. The remains in Ntarama church were fresh in our minds.

We visited the Amahoro group of mountain gorillas. Amahoro means "peace" in Kinyarwandan. We spent the morning with the large, gentle leader, a silverback named Ubumwe, meaning "unity," and met a recent young addition to the group, a female gorilla named Rwanda. Rwanda is in Peace led by Unity.

Anyway that is what we hoped for. To some extent, Rwanda seemed to be moving in that direction. Rwandans told us, "You can walk late in the night; no one will touch you, and no one will hurt you." We always felt safe, even though we knew we were standing on the same soil that had absorbed the blood of massive atrocities just a decade earlier. A young boy we met told us, "In Rwanda it can never happen again." I could feel some weird duality: pushing forward but not leaving loss behind.

Years later, I befriended a young man from Cote d'Ivoire who had been forcibly recruited as a child soldier, yet had managed to escape. He asked me if I'd been to Africa. I told him about my trips to Cameroon and Rwanda. I showed him pictures of the chimpanzees orphaned by the bushmeat trade, whom I had cared for. He gazed at the baby girls sitting in my lap sucking on milk bottles.

"People eat them?" he asked, and without waiting for a reply, added, "They are human!"

I showed him pictures of the giant gorillas of Rwanda.

"Did you smoke something before you went to see them?" he joked. They looked dangerous to him, and he wondered why I was not afraid.

He had been forced to fight in his nation's civil war. When I mentioned Rwanda, he told me he never wanted to go there. He only said, "Those people killed like crazy."

We watched a video slideshow I had put together from my trip. When we got to the images from the Ntarama church memorial, the camera became shaky. My filming hands had been unsteady, and my slight sobs were audible in the background. Looking at the photos and all the skulls, he asked why the bodies had been kept on display.

"I guess as proof that it really happened."

"No. They should bury those bodies," he insisted. After a pause, he added, "They won't forget."

There would be reminders.

Their hands reminded me. In that restaurant on our last night in Rwanda, I kept staring at those men's hands, unsure of the sign language they were using. In Butare I had met a boy named Felix who used French Sign Language, which is linguistically very similar to ASL. I learned both ASL and French so I could be in the company of primates. My signing was better than my French, and now I wanted to join this restaurant conversation. I planned my entry like an eager girl jumping into a game of double dutch.

The men caught my gaze and laughed. A young man came over to our table and told us the deaf men were saying how beautiful we were. The sign they used for "beautiful" looked like "wonderful" to me.

"Oh, really?" I asked with some suspicion. "How do you know what they are saying?"

"My mother and brother are deaf," he said.

He sat down with us to finish his meal. The restaurant filled. Music blasted. People laughed. We talked about our travels. The young man was from Rwanda but had spent the last decade in Kenya. I asked him if his mother and brother were still in Kenya. He replied with only three words, and those words changed everything.

"No. Dead. Genocide."

I'm sure the music did not stop. The laughter in the room did not cease. But I heard nothing.

Two pairs of hands signing. Two pairs of hands silenced.

I am not a primatologist, nor a psychologist, but I have seen a chimpanzee spin her head in figure eights, and I have felt little orphaned fingers clinging to my shirt. I have seen skulls lined up on a shelf in a wooden church in Ntarama, and I have witnessed the way we deal with hazardous wastes in the United States. I know our human hands are soiled, and I wonder how we can clean up our messes.

What I know about resilience and resistance, I learned from studying soils. When loaded heavily or too quickly, soils undergo a bearing-capacity failure—they collapse. Even after heavy loads are removed, the memories of previous burdens remain in each grain of clay. Interestingly, over time soils can gain strength from withstanding such pressures.

I think about this as an engineer, a human being, a primate. Recovery requires reconnaissance. We must all understand the legacy of soiled human hands, and we must keep in mind that where we go from here is also in our hands.

Opportunities to Work with Primates

For the latest information on working with nonhuman primates, please visit Primate Watch (http://www.primatewatch.org/). This appendix lists a handful of recommended organizations working in education and outreach, antivivisection, noninvasive research, and rescue and rehabilitation, as well as a number of sanctuaries.

Education, Outreach, Conservation Organizations

Friends of the Earth Malaysia (FOEM, or Sahabat Alam Malaysia, SAM) (http://www.foei.org/); general advocacy; Penang, Malaysia, Southeast Asia

International Primate Protection League (IPPL) (http://www.ippl.org/; http://www.ippl-uk.org/); general advocacy; volunteer program; Summerville, South Carolina, USA, and London, UK

Neotropical Primate Conservation (http://www.neoprimate.org/); general advocacy with strong environmental focus; La Esperanza, Peru, South America

Organizations Working against Research Using Nonhuman Primates

In Defense of Animals (IDA) (http://www.idausa.org/about.html); general advocacy with a strong antivivisection focus; San Rafael, California, USA

People for the Ethical Treatment of Animals (PETA) (http://www.peta.org/); general advocacy with a strong antivivisection focus; Norfolk, Virginia, USA

Physicians Committee for Responsible Medicine (PCRM) (http://www.pcrm.org/); specializing in noninvasive research and strongly opposed to animal exploitation; Washington, DC, USA

Stop Animal Exploitation NOW! (SAEN) (http://www.all-creatures.org/saen/); focused on fighting animal experimentation; Milford, Ohio, USA

Organizations Conducting Noninvasive Primate Studies

Chimpanzee and Human Communication Institute (http://www.cwu.edu/~cwuchci/); communication; volunteer program; Ellensburg, Washington, USA

Mona Foundation (http://www.fundacionmona.org/en/); mixed species; volunteer program; Riudellots de la Selva, Spain

Physicians Committee for Responsible Medicine (PCRM) (http://www.pcrm.org); research on cancer, diabetes, and human health; Washington, DC, USA

Wildlife Friends of Thailand (WFFT)
(http://www.wfft.org/); volunteer program; Phetchaburi,
Thailand, Southeast Asia

Rescue and Rehabilitation Organizations

Gibbon Rehabilitation Project (GRP)
(http://www.gibbonproject.org/); gibbons; volunteer
program; Phuket, Thailand, Southeast Asia

Neotropical Primate Conservation (http://www.neoprimate.
org); volunteer program; La Esperanza, Peru, South America

Wildlife Friends of Thailand (WFFT)
(http://www.wfft.org/); volunteer program; Phetchaburi,
Thailand, Southeast Asia

Sanctuaries

Centre for Animal Rehabilitation and Education (CARE)
(http://www.primatecare.org.za); baboons; volunteer program; Phalaborwa, South Africa

Chimps Inc. (http://www.chimps-inc.org/); chimpanzees;
volunteer and internship program; Bend, Oregon, USA

Foundation AAP Sanctuary for Exotic Animals
(http://www.aap.nl/english); mixed species; volunteer
program; Almere, Netherlands

Gibbon Rehabilitation Project (GRP) (http://www.gibbonproject.org/); gibbons; volunteer program; Phuket, Thailand, Southeast Asia.

Highland Farm (http://www.highland-farm.org/); gibbons; volunteer program; Tak, Thailand, Southeast Asia

Ikamaperu (http://www.ikamaperu.org); monkeys; volunteer program; Tarangue, Peru, South America

International Animal Rescue (IAR) Indonesia (http://www.internationalanimalrescue.org); macaques and lorises; Bogor, Indonesia, Southeast Asia.

International Primate Protection League (IPPL) (http://www.ippl.org/); gibbons; volunteer program; Summerville, South Carolina, USA

Inti Wara Yassi (http://www.intiwarayassi.org/articles/volunteer_animal_refuge/home.html); volunteer program; Bolivia

Jungle Friends Primate Sanctuary (http://www.junglefriends.org/); monkeys; volunteer and internship program; Gainesville, Florida, USA

Mona Foundation (http://www.fundacionmona.org/en/); mixed species; volunteer program; Riudellots de la Selva, Spain

The Monkey Sanctuary (http://www.monkeysanctuary.org/); monkeys; volunteer program; Looe, Cornwall, UK

Talkin' Monkeys Project, Inc. (http://www.talkinmonkeysproject.org); mixed species; volunteer program; Clewiston, Florida, USA

Pan-African Sanctuaries Alliance (PASA) (http://pasaprimates.org/); various organizations working with nonhuman primates across Africa.

Wildlife Friends of Thailand (WFFT) (http://www.wfft.org/); volunteer program; Phetchaburi, Thailand, Southeast Asia

Contributors

Michael A. Budkie, AHT, is the executive director of Stop Animal Exploitation NOW! (SAEN at http://www.saenonline.org), which he cofounded to focus on ending animal experimentation. As a student at the University of Cincinnati, Budkie witnessed the atrocities of animal experimentation and, as a result, campaigned and stopped a cat head-injury experiment there, launching his career as an animal activist. SAEN has terminated research projects at several universities and forced the USDA to take legal action against multiple laboratories. In his twenty years of activism, Budkie has appeared on numerous TV and radio programs and published articles and books to expose the truth about animal experimentation.

British zoologist Keri Cairns began working with primates in 1998 at a monkey sanctuary in Cornwall, where he remained for nine years. It was here that he first came across woolly monkeys and capuchins. Cairns then moved temporarily to Ireland to manage a monkey sanctuary, where he tended ex-lab capuchins, an ex-circus Japanese macaque, and an old De Brazza's monkey. He currently works with rescued shire horses at Chawton House Library (http://www.chawton. org/), a working estate that used to belong to Jane Austen's brother, Edward Austen Knight. He continues to help the International Primate Protection League (IPPL) as needed, whether traveling to Thailand to work with gibbons, or to

Gibraltar to photograph Barbary macaques. He maintains two websites: http://
kericairns.blogspot.com/ and http://picasaweb.google.com/KeriCairns

Until her retirement in 2000, BARBARA G. COX was a professional science
writer. For more than forty years, she researched and wrote papers for scholarly
journals and published books and articles on health topics. As a sideline, she volunteered her services to nonprofit animal-welfare organizations, writing feature
stories, newsletters, and brochures, and that has become her primary focus. Cox
lives within a short drive of Jungle Friends Primate Sanctuary, where she often
visits the two capuchin monkeys she sponsors: Cappy and Puchi.

DEBRA DURHAM is an ethologist who specializes in the study of animal responses to change and trauma and nonhuman primate behavior. Over the years,
she has worked with primates in a range of settings, including studying macaques
in laboratories and endangered lemurs in the wild. More recently, Durham's focus has been on pain and suffering caused by captivity, especially for scientific
experimentation. She currently works for the Physicians Committee for Responsible Medicine (PCRM) (http://www.pcrm.org), where she studies psychological conditions in primates and other animals and strives to improve research
regulations and ethical standards.

SANGAMITHRA IYER is a plant-eating primate who is humbled to have been in
the presence of great apes. She is a licensed professional civil engineer who served
as the assistant editor of Satya magazine, a publication dedicated to animal advocacy, environmentalism, and social justice. She has volunteered at primate rescue
and rehabilitation sanctuaries in the United States and Africa. Iyer is an associate at Brighter Green (http://www.brightergreen.org), a public-policy action
organization focused on equity, sustainability, and rights, and is researching the
globalization of industrial animal agriculture. She is working on her first book,
which blends memoir, family history, and reportage to explore engineering, activism, and social movements.

PHAIK KEE LIM has been an animal advocate for twenty-five years. She lives
and works in Penang, Malaysia, for Friends of the Earth Malaysia (FOEM, or

Sahabat Alam Malaysia, SAM, at http://www.foe-malaysia.org.my). Lim writes letters to the media, corresponds with dignitaries, and educates the public—all for the benefit of wilderness and wildlife. She works to protect primates from hunters and those who would take them from their jungle homes for paltry human amusement, locally or abroad.

SHIRLEY MCGREAL founded the International Primate Protection League (IPPL) in 1973 (http://www.ippl.org) with the specific aim of protecting and conserving all species of primates, from the pygmy mouse lemur to the mountain gorilla. As IPPL's chair, she protects primates from what has become a world-wide, rampant, yet largely illegal, trade by exposing offenders. IPPL also rescues animals victimized by this trade and runs a sanctuary for gibbon apes. IPPL supports many primate sanctuaries in Africa, Asia, and South America and helps fight the abuse of captive primates. McGreal recognizes that all primates are equal—from lesser-known monkeys to the trendier great apes. Her philosophy is that "all primates are equally capable of suffering, and they are equally deserving of compassion."

Australian-born FIONA MIKOWSKI worked at the Gibbon Rehabilitation Project in Phuket, Thailand. She holds a bachelor's degree in animal science (2008) from La Trobe University in Bundoora, Australia, and has volunteered at several animal projects, zoological parks, and sanctuaries. She enjoys traveling and looks forward to a career working with animals who have previously been exploited for entertainment or otherwise misused and abused and need lifelong care and rehabilitation.

RITA MILJO grew up in war-torn Germany, then moved to South Africa in the 1950s, where her love for wildlife became apparent. For many years, she assisted urban mammals and garden birds, but her life took a new direction when she rescued a baby baboon, Bobby. When she discovered that there were no facilities to nurture such orphaned primates, Miljo founded CARE (Centre for Animal Rehabilitation and Education at http://www.friendsofcare.com) in 1989. CARE specializes in the rescue, care, rehabilitation, and release of chacma baboons in Phalaborwa, South Africa.

DEBORAH D. MISOTTI has worked with nonhuman primates for many years with a particular emphasis on gibbons. Married for more than forty years to Thomas (a LEED—Leadership in Energy and Environmental Design—accredited professional), she is the cofounder and director of the Talkin' Monkeys Project, Inc. (http://www.talkinmonkeysproject.org), a 501(c)(3) educational primate sanctuary in southwestern Florida. This environmentally conscious rescue project is the logical culmination of a lifetime of volunteering to help nonhumans, youth at risk, and the larger community, based on the couple's shared desire to make a difference. They are longstanding members of the IPPL.

PAULA MUELLNER attended Long Island University's Southampton College, where she received a bachelor of science degree in environmental science and a bachelor of arts degree in biology. She worked for many years with threatened- and endangered-species programs in county, state, and national parks. In 2002 Muellner switched to the nonprofit world and began to work with nonhuman primates. She became a sanctuary manager in Texas, working with various monkey species, and in 2006 received her master of science degree in ecology from Shippensburg University in Pennsylvania. In 2007 she became the executive director of Chimps Inc. (http://www.chimps-inc.org), a sanctuary in Bend, Oregon. She is a political and social advocate for primates (and all animals) and tries to educate people about chimpanzee conservation.

Due to escalating political conflicts, JUAN PABLO PEREA-RODRIGUEZ and his family left Colombia in 2001. As an undergraduate at Florida International University, Perea-Rodriguez was interested in the way animals (especially primates) cope with changing environments. He worked with owl monkeys at the DuMond Conservancy for Primates and Tropical Forests in Miami and the owl monkey project in the wilds of Formosa, Argentina, where he studied behavior and parasitology using noninvasive methods in the hope of helping conserve this protected species. Perea-Rodriguez now helps to protect the Florida Everglades ecosystems at an ecology lab at Florida International University.

MATT ROSSELL spent two years working undercover as a primate technician at Oregon Health and Science University's Oregon Regional Primate Research Center and is currently the campaigns director of Animal Defenders

International (ADI at http://www.ad-international.org/). His work as an undercover investigator exposed Rossell to the horrors of abuse inside research labs, factory farms, circuses, fur farms, and a slaughterhouse in his hometown of Omaha, Nebraska. Rossell is an ethical vegan and lives in Los Angeles, California, with his wife, daughter, and a menagerie of rescued dogs.

KARMELE LLANO SANCHEZ of Spain has worked with primates in Venezuela, Holland, and Indonesia. Inspired by capuchin and howler monkeys at a rescue project in Venezuela, Sanchez earned a veterinarian bachelor's degree, then headed for Holland to work with primates rescued from circuses, illegal pet ownership or trade, and laboratories. Eager to tackle wildlife trade "in situ" and return animals to their natural homes, Sanchez traveled to Indonesia, where she ultimately founded International Animal Rescue (IAR) Indonesia (http://www.internationalanimalrescue.org/indonesia/), the country's first rescue and rehabilitation facility for macaques and lorises. Currently Sanchez is pursuing a master's degree in veterinary conservation medicine in Australia to facilitate her conservation work.

NOGA AND SAM SHANEE have considerable experience in primate conservation, animal rescue, and reintroduction in South America, Asia, and the Middle East. Hailing from Israel and England, respectively, they met while volunteering in Bolivia in 2001 and were married while working at the Gibbon Rehabilitation Project (GRP) in Phuket, Thailand. Their ceremony was the first gibbon-style wedding, complete with the bride swinging down from the trees. Noga and Sam each have a master of science degree in primate conservation, and Noga is currently pursuing a PhD in political ecology. In 2007, together with friends, they formed Neotropical Primate Conservation (http://www.neoprimate.org), an NGO to take on conservation projects in South America.

BIRGITH SLOTH is a Danish biologist working in nature and species conservation and management with thirty-one years of experience. Sloth was head of the office of Convention on International Trade in Endangered Species(CITES) at the Danish Ministry of Environment for seventeen years, implementing and enforcing its provisions in Denmark and throughout the European Union. To help combat illegal trade in endangered species, she ran public-awareness campaigns and trained customs and police officers. In 1994 Sloth took a job as a consultant—still specializing in CITES—traveling to eastern and central Europe, Russia, China, Thailand, South Africa, the Middle East, and the Caribbean. In the course of her work, she has visited eighty-two countries.

HELEN THIRLWAY first worked with primates at the Monkey Sanctuary in Cornwall, England, where she helped care for a colony of Amazonian woolly monkeys and a group of rescued ex-pet capuchins and campaigned to end the United Kingdom trade in primates as pets. In Ireland, standing in for the manager on vacation, Thirlway cared for twenty-three capuchins, two Japanese macaques, and one De Brazza's monkey. She now runs the U.K. branch of IPPL.

LINDA D. WOLFE is chair of the Department of Anthropology at East Carolina University in Greenville, North Carolina. Wolfe earned her PhD in 1976 from the University of Oregon in anthropology and field primatology. She has studied wild Asian monkeys (macaques) in Japan, India, Bali, and Indonesia as well as semi-free-ranging, translocated troops of macaques in the United States. She is currently working to establish ethnoprimatology, a new field of study focusing on the interaction between human and nonhuman primates.

References

"Asian Animals: Tarsier." Available online at http://library.thinkquest.org/5053/
Asia/tarsier.html.

"Awards and Congratulations!" 2008. *IPPL: International Primate Protection
League News* 35 (1) (May): 3.

Brown, Lester R. 2008. *Plan B 3.0: Mobilizing to Save Civilization*. New York:
W. W. Norton and Company.

Brown, Paul. 1999. "Malaria Vaccine Spells Hope for Millions." *Guardian*,
July 28. Available online at http://www.guardian.co.uk/science/1999/jul/28/
infectiousdiseases.guardianweekly (accessed January 30, 2008).

Burbacher, Thomas M., and Kimberly S. Grant. 2000. "Methods for Studying
Nonhuman Primates in Neurobehavioral Toxicology and Teratology."
Neurotoxicology and Teratology 22 (4) (July–August): 475–86.

Cairns, Keri. 2008. "News from Highland Farm." *IPPL: International Primate
Protection League News* 35 (3) (December): 16–17.

"Chimpanzee Research Facts." In Defense of Animals. Available online at http://
www.idausa.org/facts/chimpresearch.html (accessed August 27, 2009).

"The Chinese Monkey Connection." 2008. *IPPL: International Primate
Protection League News* 35 (2) (September): 11.

"Chinese Report on Trade in Crab-Eating Macaques." 2009. *IPPL: International
Primate Protection League News* 36 (1) (May): 12–13.

"The CITES Treaty and the International Primate Trade." 2008. *IPPL: International Primate Protection League News* 35 (1) (May): 8–9.

Convention on International Trade in Endangered Species of Wild Fauna and Flora (CITES). 1979. Available online at http://www.cites.org/eng/disc/text. shtml (accessed August 11, 2009).

Crair, Ben. 2008. "The Forgotten Ape: Why Can't the Gibbon Get Any Respect?" *IPPL: International Primate Protection League News* 35 (2) (September): 6–7.

"European Parliament: Ban Primate Experimentation!" 2007. *IPPL: International Primate Protection League News* 34 (3) (December): 22.

Fleagle, John G. 1999. *Primate Adaptation and Evolution.* 2d ed. San Diego: Academic Press.

Food and Agriculture Organization of the United Nations (FAO). 2006. Livestock's Long Shadow: Environmental Issues and Options. Available online at ftp://ftp.fao.org/docrep/fao/010/a0701e/a0701e00.pdf.

"The Future for Lab Primates." 2008. *IPPL: International Primate Protection League News* 35 (3) (December): 14.

Groves, Colin. 2001. *Primate Taxonomy.* Washington, DC: Smithsonian Institution Press.

———. 2008. "Introducing the Peculiar Proboscis Monkey." *IPPL: International Primate Protection League News* 35 (3) (December): 10–12.

Hance, Jeremy. 2009. "Lessons from the Crisis in Madagascar: An Interview with Erik Patel." August 11. Available online at http://www.illegal-logging.info/ item_single.php?it_id=news&printer=1.

Herman, Judith. 1992. *Trauma and Recovery: The Aftermath of Violence from Domestic Abuse to Political Terror.* New York: Basic Books.

"Highland Farm and the Plight of Thailand's Gibbons." 2009. *IPPL: International Primate Protection League News* 36 (1) (May): 6.

Honigsbaum, Mark. 2001. "The Monkey Puzzle." *Guardian*, November 24. Available online at http://www.guardian.co.uk/education/2001/nov/24/ research.highereducation (accessed January 30, 2008).

"Hundreds of Monkeys Confiscated in Malaysia." 2007. *IPPL: International Primate Protection League News* 34 (2) (September): 12.

International Union for Conservation of Nature. Red Data List. Available online at http://www.iucnredlist/org/.

"IPPL Members' Meeting 2006: Conference Brings Primate Protectors Together in South Carolina." 2006. *IPPL: International Primate Protection League News* 33 (1) (June): 3–10.

Masserman, Jules H., Stanley Wechkin, and William Terris. 1964. "'Altruistic' Behavior in Rhesus Monkeys." *American Journal of Psychiatry* 121 (6): 584–85.

McGreal, Shirley. 2008. "Primate Congress in Scotland." *IPPL: International Primate Protection League News* 35 (2) (September): 7–9.

"Milgram Experiment." *Wikipedia: The Free Encyclopedia.* Available online at http://en.wikipedia.org/wiki/Milgram_experiment? (accessed October 11, 2009).

Milgram, Stanley. 1963. "Behavioral Study of Obedience." *Journal of Abnormal and Social Psychology* 67: 371–78.

Nguyen Vinh Thanh. 2009. "Delacour's Langurs: On the Brink." *IPPL: International Primate Protection League News* 36 (1) (May): 3–5.

Nursahid, Rosek. 2008. "ProFauna's New Javan Langur Rescue Center." *IPPL: International Primate Protection League News* 35 (3) (December): 18.

"Pet Chimp Attacks Owner's Friend." 2009. *IPPL: International Primate Protection League News* 36 (1) (May): 21–22.

"A Promising Proposal for Wild Non-human Primates." 2009. *AWI Quarterly* 58 (1) (Winter): 9.

"Pygmy Tarsiers Back from 'Extinction.'" 2009. *AWI Quarterly* 58 (1) (Winter): 8.

Ramos, Serafin N. Jr. "Yes, There Are Tarsiers in Sarangani." Available online at http://www.sarangani.gov.ph/tarsier.php.

Regan, Tom. 1985. "The Philosophy of Animal Rights." Available online at http://www.cultureandanimals.org/pop1.html.

———. 2003. *Animal Rights, Human Wrongs: An Introduction to Moral Philosophy.* New York: Rowman & Littlefield Publishers, Inc.

———. 2005. *Empty Cages: Facing the Challenge of Animal Rights.* New York: Rowman & Littlefield Publishers, Inc.

Reynolds, Vernon. 1971. *The Apes.* New York: Harper Colophon Books.

Sagan, Carl, and Ann Druyan. 1992. *Shadows of Forgotten Ancestors: A Search for Who We Are.* New York: Random House.

Shumaker, Robert W., and Benjamin B Beck. 2003. *Primates in Question.* Washington, DC: Smithsonian Books.

"Slow Lorises Receive International Trade Protections." 2007. *IPPL: International Primate Protection League News* 34 (2) (September): 15.

Smith, David. 2009. "The Study of Animal Metacognition." *Trends in Cognitive Science* 13 (9) (September): 389–96.

"Stanford Prison Experiment." *Wikipedia: The Free Encyclopedia*. Available online at http://en.wikipedia.org/wiki/Stanford_prison_experiment (accessed October 11, 2009).

Stevenson, Pablo R. 2000. "Seed Dispersal by Woolly Monkeys (*Lagothrix lagothricha*) at Tinigua National Park, Colombia: Dispersal Distance, Germination Rates, and Dispersal Quantity." *American Journal of Primatology* 50 (4): 275–89.

Strier, Karen. 2010. *Primate Behaivoral Ecology*. 4th ed. Boston: Prentice Hall.

Thirlway, Helen. 2009a. "Monkey Wars in Europe." *IPPL: International Primate Protection League News* 36 (1) (May): 24.

———. 2009b. "U.S. and UK Primate Imports." *IPPL: International Primate Protection League News* 36 (1) (May): 14–16.

"Undercover in a Primate Lab." 2008. *IPPL: International Primate Protection League News* 35 (1) (May): 9.

"U.S. Zoos Import Wild-Caught Monkeys." 2006. *IPPL: International Primate Protection League News* 33 (1) (June): 20–21.

"Vet Describes the Plight of Indonesia's Primates." 2008. *IPPL: International Primate Protection League News* 35 (1) (May): 7–8.

"The War on Animals: PCRM Confronts the Military's Deadly Use of Animals for Medical Training." 2009. *Good Medicine* 18 (1) (Winter): 6–7.

Weinberg, Jessica. 2007. "The Primates of Africa's Bioko Island: IPPL Helps Sponsor Awareness Campaign." *IPPL: International Primate Protection League News* 34 (3) (December): 12–14.

"Wildlife Friends of Thailand: Gibbons and More." 2009. *IPPL: International Primate Protection League News* 36 (1) (May): 9.

Williams, Oscar. 1952. *Immortal Poems of the English Language*. New York: Washington Square Press.

Woods, Vanessa. 2008. "Bonobo Paradise." *IPPL: International Primate Protection League News* 35 (3) (December): 3–5.

Index